CELLULAR IN VITRO TESTING

CELLULAR IN VITRO TESTING

METHODS AND PROTOCOLS

edited by
John W. Haycock
Arti Ahluwalia
John M. Wilkinson

PAN STANFORD PUBLISHING

Published by

Pan Stanford Publishing Pte. Ltd.
Penthouse Level, Suntec Tower 3
8 Temasek Boulevard
Singapore 038988

Email: editorial@panstanford.com
Web: www.panstanford.com

British Library Cataloguing-in-Publication Data
A catalogue record for this book is available from the British Library.

Cellular in vitro Testing: Methods and Protocols

ISBN 978-981-4364-97-3 (Hardcover)
ISBN 978-981-4364-98-0 (eBook)

Printed in the USA

Contents

Preface

The laboratory culture of eukaryotic cells has now been routine practice for over four decades, underpinning a multitude of biological and medical applications—from cancer studies to drug discovery, from toxicology to stem and developmental cell biology. Irrespective of the applications, the tools and techniques that are used to grow an adherent cell in vitro outside a living organism comprise of the cell culture medium, the tissue culture incubator and a substrate, most commonly made from "tissue culture polystyrene" or glass. The substrate, in almost all instances, takes the form of a flat two-dimensional (2D) surface. Although a very large number of biological studies have been performed using 2D in vitro cell culture, one might be forgiven for questioning the relevance of such studies when interpreting the information in light of in vivo models, which consider the relevance of multi-cellular tissues and organs, a blood supply, an immune system, plus endocrine and neuronal signals.

To overcome the limitations of 2D models, a number of 3D in vitro models are being developed for a range of tissues and organs, which take account of the spatial organisation and arrangement of the cells therein. The development of all such models has a common objective, which is to establish scientifically robust platforms for enabling the improved interpretation of data beyond what simple 2D cell cultures provide. A more ambitious objective is to approach the physiological relevance of an in vivo model—and furthermore, in so doing to **R**eplace, **R**educe, or **R**efine the necessity of the in vivo model itself, especially those based on animal studies. Thus, the development of 3D cell culture in vitro models directly addresses the 3Rs principle, first introduced by Russell and Burch in 1959 in their book *The Principles of Humane Experimental Technique.*

For this to become a possibility, it is necessary to establish a growth environment that mimics the native tissue structure as closely as possible. Thus, 3D culture models typically combine materials science, cell biology, and bioreactor design. The integration of these approaches is particularly important, given the practical and applied directions of such work. For example, the promise of regenerative medicine to replace body parts suffering acute injury or degeneration associated with aging is frequently reported in the popular media. However, the routine uses of such therapeutic treatment are presently few and far between. To accelerate progress, accurate and relevant 3D culture models must be developed, as they will be essential for the development of such technologies.

3D cell culture models generally include the study of whole animals and organotypic explant cultures (including embryos), cell spheroids, microcarrier cultures, and tissue-engineered constructs. So while not all 3D cell culture models require a scaffold, their use has seen a rapid increase in recent years. This is in large part due to advances in biomaterial science and fabrication methods for 3D printing, for example, micro-stereolithography of polymers for creating scaffolds with micrometer resolution, or similarly electrospinning, micro-extrusion, micro-injection moulding and ink jet printing.

The potential to use human primary cells or stem cells is highly relevant when developing 3D in vitro cultures, not only for considering the 3Rs as an alternative approach but also for understanding the fundamental processes of cellular differentiation. This becomes particularly important when considering methods for therapeutic intervention. The ability to isolate cells with the capacity to renew, mitotically divide, and differentiate into a diverse range of cell types is of fundamental importance—and yet many studies on stem cells still use a 2D environment. New evidence is now emerging on the relevance of the 3D environment and the ability of cells to "sense" their 3D environment including the extracellular matrix stiffness, reported to determine stem cell differentiation along a particular lineage. This logically extends, though, to the culture environment and the ability to supply nutrients in the culture medium to a growing niche of metabolically active cellswhile simultaneously removing the waste products for maintaining survival. Considerable evidence also exists on the development and need for communication of different cell types in 3D culture, where paracrine signalling is responsible for the differentiation and development of cell types in co-culture to form a nascent functioning tissue, for example, keratinocytes and fibroblasts for 3D skin models or neuronal and glial cells for 3D nerve models. Thus, the rapid development of bioreactor systems for allowing controlled flow, the spatial organization of cell–scaffold constructs, as well as organ crosstalk for this purpose is gaining momentum.

Cellular in vitro Testing: Methods and Protocols reports on a wide range of methods for the applications of 3D in vitro cell culture models, either for the purposes of in vitro testing or in the long-term development of forming tissues for therapeutic purposes. The scope and contents of the book have arisen from selected presentations given at the Annual *Quasi-Vivo*® User Group Meetings, and the style of each chapter is based on a descriptive protocol style, such that readers will be able to reproduce individual methods in the laboratory step-by-step. The book starts with a review chapter, which gives an overview of methods for connected culture experiments using perfused flow chambers. Thereafter, there are nine chapters which cover key areas in human airway inhalation toxicity, quantifying xenobiotic metabolizing enzymes in hepatocytes, cardiac patches, risk assessment of nanoparticles, the development of nanosensor scaffolds, a paracrine human skin model for irritancy detection, drug testing using intestinal mucosa, a human bronchial model and the use of a porous scaffold for simple and routine 3D culture.

We hope this book serves as a useful methods and protocols manual for laboratory scientists who need to develop the underpinning scientific basis and technical details for 3D in vitro cell cultures. We are personally indebted to all of the international experts who have kindly contributed chapters and taken great care and time in preparing their contributions for this book.

John W. Haycock
Arti Ahluwalia
John M. Wilkinson
Summer 2014

Chapter 1

Methods for Conducting Connected Culture Experiments Using the Quasi-Vivo® Chambers

Tommaso Sbrana and Arti Ahluwalia

Interdepartmental Research Center "E.Piaggio", Faculty of Engineering, University of Pisa, Via Diotisalvi, 2 - 56126 Pisa, Italy

arti.ahluwalia@centropiaggio.unipi.it

In the human body the vascular network links organs, which, besides transporting oxygen and nutrients, is a vital communication highway for metabolic signaling between tissues. Unraveling the complexity of interorgan and intertissue cross talk in vivo is a complex and challenging task. Intelligent in vitro models able to recapitulate the physiological interactions between tissues in the body connected by the bloodstream have enormous potential as they enable detailed studies on specific two-way or higher-order organ–organ and tissue interactions. An integrated systemic view can thus be constructed piecewise. The aim of this paper is to show how allometry can be used to set up physiologically relevant connected culture models of biotransformation, distribution, adsorption and gas exchange in the Quasi-Vivo® system.

Cellular in vitro Testing: Methods and Protocols

Edited by John W. Haycock, Arti Ahluwalia, and John M. Wilkinson

ISBN 978-981-4364-97-3 (Hardcover), 978-981-4364-98-0 (eBook)

www.panstanford.com

1.1 Introduction

Much of what we know of integrative physiology comes from animal studies, many of which currently rely on knockout animals and induced disease models. Translating the results to humans is not straightforward, and as several studies have shown, caution needs to be exercised when extrapolating results to human physiology pathology or toxicity [1]. Moreover animal models are not amenable to decomposition of biomolecule dynamics in different tissues and organs. As a consequence identifying the contribution of individual tissues or organs to the absorption, distribution, metabolism and excretion (ADME) of a drug, chemical or metabolite is a formidable task. On the other hand, although there is clearly a network of signaling between different tissues which contributes to maintaining homeostasis in the human body, in vitro research continues to focus on alterations in cell signaling at the membrane or cytoplasmic level. Thus much of our understanding of signal transduction is limited to a very small space and time window. The question of how signals are propagated to and translated by remote tissues and organs, and how they modulate the internal milieu has not yet been addressed. Indeed very little research has been carried out on higher-level models of metabolism or toxicity containing different cell or tissue types.

Although in vitro models have evolved to a certain degree, most laboratories still use the classic petri dish and rely on acute experiments to study cell response to a variety of externally applied stimuli (toxins, growth factors, adhesion ligands). Generally in drug or chemical toxicity studies, cell or tissue cross talk is simulated using growth factors and hormones or conditioned media, for example to induce differentiation or to maintain cell phenotype. Cocultures are an alternative option, increasingly common in three-dimensional (3D) tissue-engineered constructs, but they are difficult to characterize particularly if the different cell types are not spatially separated. Moreover to have any relevance to the physiological context, contact cocultures in petri dishes must necessarily involve cell types that reside together in vivo. In fact most coculture studies combine endothelial cells or fibroblasts with parenchymal cells, usually in layers, to probe functional expression and cell vitality, rather than to probe cross talk between cells.

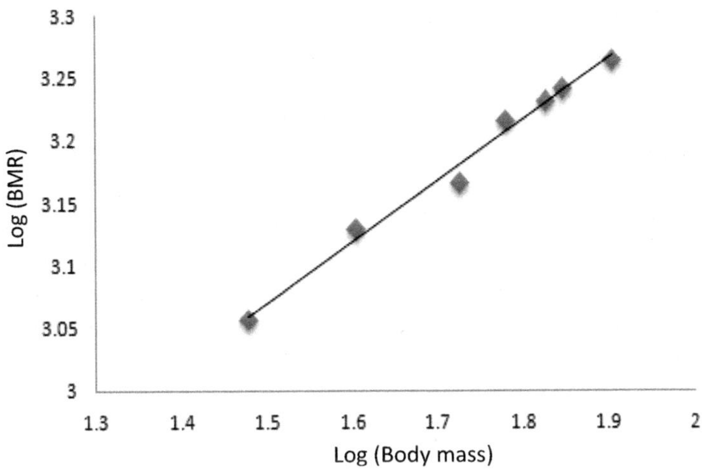

Figure 1.1 BMR correlated with the body mass of a sample population of males and females.

Recapitulating systemic pathophysiology or toxicity in vitro is a challenging task and requires a methodical engineering approach, breaking down the vascular circuitry to its most basic elements and reconstructing the most important features of the network in vitro. By increasing the number of interactions and variables stepwise in a properly scaled model, signaling dynamics between organs and their contribution to the whole-body response can be systematically investigated.

The Quasi-Vivo® system was specifically developed for systematically reconstructing tissue and organ cross talk in vitro [2]. It is a modular system consisting of interconnected chambers, each one designed to house a specific tissue or organ (Fig. 1.1). Cellular or tissue cross talk is achieved by a dynamic, flow-dependent transport of molecules from one chamber to the others, thus recapitulating both biomechanical and biochemical stimuli present in vivo. Higher-order, more complex models can thus be assembled by combining two or more cell types or tissues connected by the flow of a common medium. By analyzing the dynamics of metabolite or signaling molecule levels in different chronic or acute conditions in the system, cross talk between the tissues can be identified and probed by periodic analysis of the common medium, much as blood samples are periodically taken from patients.

To set up and scale the system it is necessary to use a methodical approach, first by selecting an appropriate subsystem to study, probing first one element of the system and then two, three and more at a time. In addition, it is necessary to use physiologically relevant cell ratios and flow rates that do not cause shear stress–related damage to cells and that allow adequate residence times in each compartment to enable cells to process biochemical signals. Here we show how the Quasi-Vivo® chambers can be used to design physiologically relevant, connected culture systems of biotransformation adsorption and gas exchange based on allometric scaling.

1.1.1 *Allometric Scaling*

Allometry is the science of scaling and deals with changes in body size and relationships amongst different parameters and processes in all organisms as a function of body mass M [3]. The basic allometric equation can be used to correlate physiological variables between organisms of different sizes:

$$Y = a * M^b \tag{1.1}$$

Y stands for a physiological parameter that has to be correlated to body mass (M), such as the basal metabolic rate (BMR), heart rate life span, etc. a is a proportionality factor, whereas b is the allometric exponent. b varies in magnitude and sign and has a specific value for each parameter according to how it scales with mass. Typically $b = 1$ for volumes and cell numbers and lies between 0 and 1 for rates (metabolic rates, flow rates, etc.), while it is negative for frequencies (cardiac frequencies, respiratory frequencies, etc).

To estimate the value of b, data on the parameter are plotted in a log-log graph in relation with body mass. b is the slope of the straight line that best fits the data. Figure 1.1 shows a log-log graph of the BMR of a sample population of adults plotted against body mass. The estimated slope is 0.75, and this is the well-known value of the allometric exponent for BMR. Like b, a varies from parameter to parameter. The value of a can be estimated from the y-intercept of the log-log graph. Alternatively if b and the mass as well as the parameter are known for one organism, a can be calculated from

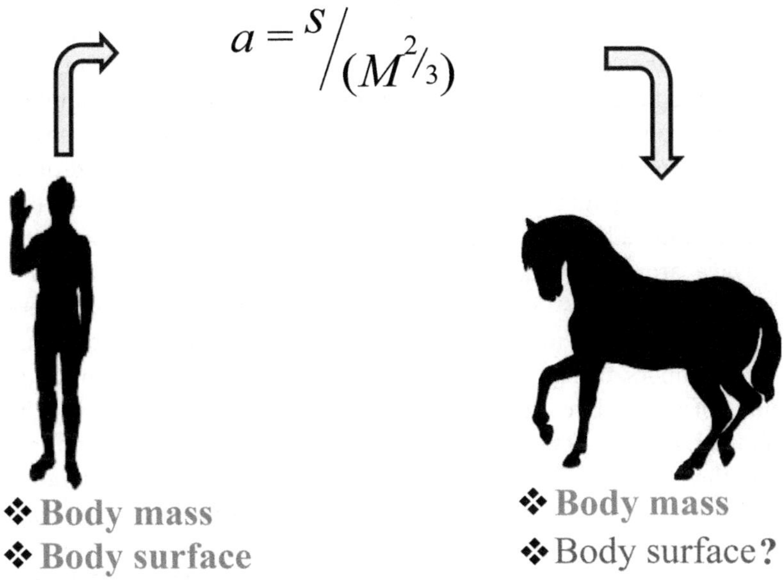

Figure 1.2 Algorithm for estimating the body surface area of a horse, given the mass and surface area of a human.

Eq. 1.1. Successively the parameter can be estimated for any animal of known mass. For example, to estimate the body surface area (S) of an animal we can use the well-known equation that links surface area to body mass. The equation is valid for all mammals:

$$S = a * M^{2/3} \quad (1.2)$$

If we wish to estimate the surface area of a 200 kg horse, given that of a 70 kg man (1.73 m^2) [4], we first find the value of a using the human data and then substitute a and the mass of the horse in Eq. 1.2. Figure 1.2 summarizes the passages to evaluate horse body surface area. The starting points are the body mass and body surface of a human.

The allometric approach can be used to evaluate the relationships between different tissues in order to ensure that the same relationships are conserved in the downscaled bioreactor environment. In theory it is possible to set up an in vitro experiment in which all the allometric relations between physiological parameters are conserved at the same time. However, the allometric approach does not take into account experimental constraints in the in vitro environment, such as dimensions of the chamber where cells are seeded, flow rates that pumps are able to impose, oxygen consumption of different cells types, etc. For example, allometric scaling of cell numbers and exchange surface areas is often incompatible in vitro because in physiological conditions tissues are organized in 3D structures, whereas most in vitro studies are conducted using monolayer cultures. Scaling the exchange surface area of an organ in order to mimic this parameter in a bioreactor implies the use of monolayer cell numbers that are not equal to the ideal scaled cell number. Likewise, estimating cell numbers using allometry will lead to monolayers, which do not correspond to the ratio between surfaces estimated by the allometric approach. In such cases 3D cultures are the best option, and the Quasi-Vivo® chambers are designed to house both scaffolds and monolayers. Alternatively, two different approaches can be used to set up a connected culture system as described here. The first approach is the metabolic and surface scaling model (MSSM). In this example described we wish to study the cross talk between liver and endothelial tissue in order to be able to set a multiorgan in vitro experiment on biotransformation and distribution. We show how liver metabolism and the endothelial surface area can be scaled in order to estimate the number of cells required for the experiments. A second model called the cell number scaling model (CNSM) simply considers cell numbers and scales them down to fit the experimental constraints.

1.1.2 *Metabolic and Surface Scaling Model*

The liver is responsible for the uptake, conversion and distribution of many of the nutrients entering the digestive tract and is also the main orchestrator of exogenous metabolism, while vascular tissue is the conduit through which signals are relayed to distant organs. We therefore begin by connecting hepatocytes with endothelial cells before adding other cells or tissues to construct an in vitro model of biotransformation and distribution. Biotransformation is a metabolic process and depends on the metabolic efficiency of cells, whereas distribution is a surface-mediated process. Therefore

Table 1.1 Summary of human data used for allometric scaling

	Values
Body mass	70 kg
BMR	88 J/s
Liver mass	1.4 kg
No. of hepatic cells	2×10^{11}

the hepatocytes in the model are scaled with reference to basal metabolism, whereas the endothelium is scaled using the surface area of the human vascular system as a starting point. As cells are usually plated in monolayers, the allometric design process begins by considering the metabolism of a two-dimensional culture of human hepatocytes in a single Quasi-Vivo® module. Hepatocytes can be seeded on the bottom surface of the chamber (either on a glass slide or on a scaffold).

Standard parameters from the human body used to establish the experimental setup in the MSSM through allometric scaling are summarized in Table 1.1.

The surface area of the cell culture zone of a Quasi-Vivo® chamber is 1.33 cm^2, so it is possible to estimate the number of hepatocytes that can be seeded in a monolayer. A confluent layer of hepatocytes has a density of about 2×10^5 cells/cm^2, which corresponds to 2.6×10^5 cells. The liver generates 27% of the total BMR of a human, corresponding to 23.76 J/s. Then, assuming the total metabolic contribution of the liver is due only to hepatocytes, the BMR per human hepatocyte is 119 pJ/s, while the equivalent BMR of the Quasi-Vivo® liver is 30 μJ/s. Despite being a metabolic organ, liver mass scales with an exponent of $b \approx 1$. The explanation usually given for this is that small organisms have a larger number of metabolic organelles (mitochondria) per cell (b for the number of mitochondria per cell is 4/3). Then the total BMR of the Quasi-Vivo® system is 111 μJ/s. Allometric equations can be used to find equivalent body mass of ($M_{\text{Quasi-Vivo}}$) of a Quasi-Vivo® system. Equations 1.3 and 1.4 summarize the steps required to estimate $M_{\text{Quasi-Vivo}}$.

$$\text{BMR}_{\text{man}} = a_{\text{BMR}} * M_{\text{man}}{}^{\frac{3}{4}} \tag{1.3}$$

$$\text{BMR}_{\text{Quasi-Vivo}} = a_{\text{BMR}} * M_{\text{Quasi-Vivo}}{}^{\frac{3}{4}} \tag{1.4}$$

$M_{\text{Quasi-Vivo}}$ in this case is 1 mg. Note that this value depends very much on the number of hepatocytes employed. $M_{\text{Quasi-Vivo}}$ can be upto 5 mg in 3D scaffolds or tissue slices. Having established this parameter, the allometric approach can be used to find a suitable surface area to simulate the vascular endothelial exchange area. The allometric equation that links the vascular surface area (S) to the body mass (M) of a mammal is [5]

$$S = a * M^{11/12}. \qquad (1.5)$$

Note that capillary area does not follow the 2/3 scaling law because it does not refer to an external area. Given that the capillary bed surface of a standard man in resting conditions is 500 m^2, it is possible to estimate the constant a from Eq. 1.5 and then find a suitable surface area to simulate the endothelium in an in vitro experiment.

$$S = a * (1\ mg)^{11/12} \qquad (1.6)$$

The surface of a chamber that represents the endothelium should be 0.33 cm^2. This is about a quarter of the cell culture area of a Quasi-Vivo® module. Thus if one chamber is used to simulate the exchange surface area of the endothelium, about three to four Quasi-Vivo® modules have to be used to represent the liver (less if the hepatocyte density is increased by seeding on a scaffold). Therefore to respect the MSSM, four hepatocyte chambers and one endothelial chamber are required. Note that the relationship is not linear due to the scaling laws used.

1.1.3 *Cell Number Scaling Model*

In this paragraph an alternative model is described to simulate the same problem—the study of cross talk between endothelial and hepatic tissue. In tissues, cell numbers play a very important role in physiologic functions. For example, the cell number is an important parameter to characterize drug filtration or absorption rates. As a consequence if the aim is to study drug passage from one organ to another by simulating tissues as monolayers in several bioreactor chambers connected in series or in parallel, the cell numbers in each culture is a key point in the experimental setup. An allometric model based on cell numbers could represent a valid alternative to the MSSM in order to simulate cross talk between tissues.

While the MSSM scaled down whole-body metabolism and the endothelium area, here we use the ratio between endothelial cells and hepatocytes in the visceral abdominal region. Vascular endothelial tissue represents 6.28% of human body mass. Liver mass is instead equal to 2% of human body mass. The abdomen represents 3.08% of body mass—most of this is the liver [4]. Assuming that vascular density is constant all over the body, it is possible to evaluate abdominal vascular tissue mass using the following equation:

$$M_{\text{abdomen vascular tissue}} = 0.0308 * (M_{\text{hepatocytes}} * 0.0628) \quad (1.7)$$

Liver weight is directly evaluated by its contribution to body mass (2%). Assuming that hepatocytes represent the main percentage of liver mass, Eq. 1.7 permits us to calculate the ratio (r) between abdomen endothelial weight and hepatic tissue weight:

$$r = \frac{M_{\text{abdomen vascular tissue}}}{M_{\text{hepatocytes}}} = 0.1 \quad (1.8)$$

where r represents the physiological ratio between the endothelial and the hepatic mass in the abdomen. To maintain this correlation in an in vitro experiment, the hepatocyte number has to be 10 times the endothelial cell number. This ratio can be used as a guideline to set up an in vitro experiment using one bioreactor as the hepatic module and one as the endothelial module, characterized by suitable cell numbers.

The main difference between the two models is that the MSSM gives more importance to hepatic metabolism than the CSSM. Were we to evaluate the chamber ratios using the MSSM for just the abdomen, we would require a large number of hepatocyte chambers with respect to endothelial chambers because of the difference in the allometric exponent for metabolism (3/4) and capillary area (11/12).

1.1.4 *Lung: Quasi-Vivo® Gas Exchange Model*

We now turn our attention to the evaluation of the surface area of an in vitro lung in a gas exchange model. The allometric relation that links the surface area (S) to mass (M) was given in Eq. 1.2. This equation holds for organs responsible for exchange with the external environment, such as the skin, lungs, and intestines. Given that the human lung mass is 800 g and its surface area is 80 m^2 the constant a relating lung area to body mass can be estimated. Having established the total mass of our Quasi-Vivo® system, the equivalent Quasi-Vivo® lung area using the same constant can be calculated. The surface area of the chamber, which simulates the lung, is 4.7×10^{-4} m^2.

This value can be used to design an in vitro experiment based on exchange at the air–liquid interface, as described in Ref. [7]. In particular the ratio between surface areas of the liver chamber and lung chamber provides the relationship between the number of Quasi-Vivo® modules to mimic hepatic tissue and to simulate pulmonary tissue.

$$\frac{\text{Area lung module}}{\text{Area hepatic module}} =\sim 4$$

Thus, the most sensible approach is to use one liver chamber and four lung chambers in series. Similar considerations can also be used to design Quasi-Vivo® skin and intestinal adsorption, glucose metabolism, excretion, and nutrient transport. Some examples are shown in Table 1.2.

Table 1.2 A few examples of allometric relations that have to be maintained to conserve physiological relationships between organs

Quasi-Vivo physiological model	Cells	Modules	Relationships
Biotransformation	Hepatocytes	1 liver	
Absorption and Biotransformation	Hepatocytes, Skin epithelia	12 liver, 1 skin epithelia (confluent layer)	Slin mass, $b = 0.87$, (4–8 kg) Skin area, $b = 2/3$ (1.8 m^2)
Nutrient absorption and Biotransformation	Hepatocytes, Intestinal epithelial cells	1 liver, 9 intestinal epithelia (confluent layer)	Intestinal bed, $b = 2/3$ (200 m^2)

Source: Refs. [8–10].

1.1.5 *Setting Up a Quasi-Vivo®-Connected Culture Experiment*

Because the Quasi-Vivo® environment is so different from a typical cell culture experiment, it is important to ensure that the cell types used in the different chambers do not suffer from high shear stress or the use of an inadequate medium. Therefore, before starting connected culture experiments it is necessary to establish a common medium for all cells. A suitable medium is usually that used for the most sensitive cells (e.g., the medium used for primary hepatocyte cultures when investigating connected cultures of endothelial cells, adipocytes, and rat hepatocyes). Cell morphology, vitality, and metabolic function are all known to adapt to flow [11]. Adequate internal controls using only one chamber at a time are thus essential for evaluating cell viability or function as a function of flow. After studying the cells singly they should be studied in pairs and then in threes and so on, using the monoculture flow experiments and not petri dishes as controls.

1.1.6 *Flow Rates*

Given their high sensitivity to shear, hepatocytes are the flow rate–limiting cells in a Quasi-Vivo® system. The flow velocity should allow the medium to remain in contact with the cells for sufficiently long so that they can transduce external signals from cells in adjoining modules. However, if the time is too long then cells may initiate down-regulation or other processes, limiting cross talk with cells in downstream modules. The flow rate of the adult human liver is 1,000–1,500 mL/min, and the allometric exponent *b* for flow rates is $\frac{3}{4}$, giving a maximum flow rate of ~2 μL/min in the Quasi-Vivo® system described (a total passage time of about 1,000 min per module). This flow rate depends on the number of hepatocytes seeded. In the case of a 1 mg Quasi-Vivo® body mass it is rather low and, considering the oxygen consumption requirements of hepatocytes, is not sufficient for ensuring an adequate oxygen concentration at the bottom of the modules [12]. Alternatively we can choose to reproduce the mean organ perfusion time in the liver, which is about 2–5 min giving a flow rate of about 400 μL/min.

This is close to the experimentally determined optimal flow rates of 300 μL/min and 250 μL/min for hepatocytes in Refs. [4, 13, 14], respectively.

1.1.7 *Three-Dimensional Scaffolds*

One of the most important structural cues for cells is the 3D architecture of tissues. As discussed in previous paragraphs, cells in vivo are surrounded by a network of the extracellular matrix, through which both biochemical and mechanical signals are transmitted and transferred to neighboring cells. The models presented here are based on monolayer cultures simply because this is the standard method used for in vitro preparations. For this reason we have considered organ models based mainly on absorption and permeation. The Quasi-Vivo® modules are designed to house scaffolds of up to 5 mm in height, and in Ref. [12] we show how they can be used to increase the viability and metabolic activity of hepatocytes. It has to be underlined that monolayer cultures do not suffer from problems of lack of permeation of oxygen, even if a number of chambers are linked in series, because the characteristic flow time is high enough with respect to the oxygen diffusion time to ensure continuous turnover. The Graetz number, G, is used to characterize the ratio between these two transport phenomena.

$$G = t_{\text{diffusion}}/t_{\text{convection}}$$

A Graetz number $\gg$ 1 stands for very high convection rates, meaning that oxygen is transported easily through bioreactors joined in series. On the other hand if $G \ll 1$ diffusive transport is prevalent and the last chamber of the fluidic circuit could be in hypoxic conditions. This is particularly the case for cells at the very bottom of a 3D structure, which could be exposed to hypoxic conditions, particularly if several chambers with scaffolds are placed in series. In this case it the experimental conditions should be simulated using diffusion and convection in the fluidic system to estimate G. Moreover the fluidic conditions inside a module should also be studied to avoid high flows that preclude scaffold permeation by gas from the top to the bottom, causing hypoxia in the deepest part of the 3D structure. In summary, the choice of flow rate in the presence of monolayer cultures should be based on the limiting shear stress for cells. As far as 3D structures are concerned, the shear stress is less important because scaffolds act as a shield for cells seeded in pores, but oxygen becomes the limiting parameter because of the low solubility of the gas in water. In this case flows have to be imposed taking into consideration mass transport and oxygen consumption in the system.

1.2 Conclusions

In this chapter we show how a number of engineering design principles such as allometry, mass transport, and fluid dynamic considerations can be used to set up physiologically relevant, connected culture models of biotransformation, distribution, adsorption, and gas exchange in the Quasi-Vivo® system. Besides scaling the system using physiologically and metabolically relevant cell ratios according to the system and processes under investigation, it is necessary to use flow rates that do not cause shear stress–related damage to cells and that allow adequate residence times in each compartment to enable cells to process biochemical signals. If the flow rates are too slow, however, high-density cultures may suffer from inadequate oxygen levels.

Once the basic design and experimental parameters (a common medium, a base line for assessing the effects of flow on each cell or tissue to be used) are established, the Quasi-Vivo® chambers is used to recreate the physiological microenvironment in vitro. We suggest that cell cross talk be probed stepwise beginning with two-way and then higher-order organ–organ and tissue interactions so that network maps of soluble signaling can be constructed.

Acknowledgments

This work was partially funded by the European Union Seventh Framework Programme (FP7/2007–2013) under the grant agreement n° NMP4-SL-2009-228625, InLiveTox.

References

1. Krewski, D., Westphal, M., Al-Zoughool, M., Croteau M. C., Andersen M. E. (2011) New directions in toxicity testing. *Annu Rev Public Health* **32**, 161–178.
2. Mazzei, D., Guzzardi, M. A., Giusti, S., Ahluwalia, A. (2010) A low shear stress modular bioreactor for connected cell culture under high flow rates. *Biotechnol Bioeng* **106**(1), 127–137.
3. Weibel, E. R., Bacigalupe, L. D., Schmitt, B., Hoppeler, H. (2004) Allometric scaling of maximal metabolic rate in mammals: muscle aerobic capacity as determinant factor. *Resp Physiol Neurobiol* **140**, 115–132.
4. http://miss.jsc.nasa.gov.
5. Dawson, T. H. (2003) Scaling law for capillary vessels of mammals at rest and in exercise. *Proc Biol Sci* **270**(1516), 755–763.
6. Kamiya, A., Takeda, S., Shibata, M. (1987) Optimum capillary number for oxygen delivery to tissue in man. *Mat Biol* **49**(3), 351–361.
7. Mazzei, D., Giusti, S., Sbrana, T., Ahluwalia, A. (2011) Multi-compartmental modular bioreactor as innovative system for dynamic cell cultures and co-cultures. In *Bioreactors: Design, Properties and Applications*, ISBN: 978-1-62100-164-5,163–181.
8. *The Physics Factbook*, http://hypertextbook.com/facts/.
9. Lindstedt, S. L., Schaeffer, P. J. (2002) Use of allometry in predicting anatomical and physiological parameters of mammals. *Lab Animals* **36**, 1–19.
10. Sohlenius-Sternbeck, A. K. (2006) Determination of the hepatocellularity number for human, dog, rabbit, rat and mouse livers from protein concentration measurements. *Toxicol in Vitro* **20**(8), 1582–1586.
11. Vinci, B., Murphy, E., Iori, E., Marescotti, M. C. (2010) low-regulated glucose and lipid metabolism in adipose tissue, endothelial cell and hepatocyte cultures in a modular bioreactor. *Biotechnol J* **5**(6), 618–626.
12. Sbrana, T., Ahluwalia, A. (2012) Engineering Quasi Vivo in vitro organ models. *Adv Exp Med Biol* **745**, 138–153.
13. Vinci, B., et al. (2011) Modular bioreactor for primary human hepatocyte culture: medium flow stimulates expression and activity of detoxification genes. *Biotechnol J* **6**(5), 554–564.
14. Vinci, B., et al. (2010) In vitro liver model using microfabricated scaffolds in a modular bioreactor. *Biotechnol J* **5**(2), 232–241.

Chapter 2

The Use of in vitro 3D Cell Models of Human Airway Epithelia (MucilAir™) in Inhalation Toxicity

Samuel Constant, Ludovic Wiszniewski, and Song Huang
Epithelix Sàrl, 14 Chemin des Aulx, CH-1228 Plan-les-Ouates, Geneva, Switzerland
samuel.constant@epithelix.com

In vitro assessment of inhalation toxicity is an emerging and fast-growing field. Until now, only animal models were used in OECD test guidelines for inhalation toxicity testing, and therefore in vitro alternatives of human origin are urgently needed. In the present chapter, we first review the toolbox of in vitro human airway models currently available. We then focus on the applications of the human airway epithelia (MucilAir™), a 3D cell model with a long shelf life, which is used for acute, long-term, and repeated dose testing. A panel of tests relevant for inhalation toxicology is described.

2.1 Introduction

New materials development as well as the advent of the European legislation on chemicals (Registration, Evaluation, Authorisation and

---

Dedicated to the memory of Mr. Egon Naëf

Cellular in vitro Testing: Methods and Protocols
Edited by John W. Haycock, Arti Ahluwalia, and John M. Wilkinson

ISBN 978-981-4364-97-3 (Hardcover), 978-981-4364-98-0 (eBook)
www.panstanford.com

Restriction of Chemicals [REACH]) should increase substantially inhalation toxicity testing. To date, only animal models are the accepted standards effective for the Organisation for Economic Co-operation and Development (OECD) guidelines for inhalation toxicity testing (e.g., TG 403, Draft 433, Draft 436, 412, and 413). However, to apply the 3Rs principles, any new alternative methods allowing screening and targeted testing of a relevant toxicity pathway in differentiated human cells might contribute to reduce in vivo testing, as outlined by US National research Council in *Toxicity Testing in the 21st Century* [1]. In vitro toxicity testing should be built upon test models that are relevant for the species to be protected [2]. Proper test development requires well-defined cell systems that mimic in vitro the key events that are known to occur in vivo [3].

The human respiratory tract is one of the most important entry portals of toxicants such as gasses, vapors, aerosols, fibers, particles, or nanoparticles in our body. The respiratory system can be subdivided into the upper and lower respiratory tracts. The upper respiratory tract includes the nose, pharynx, and larynx. The lower respiratory tract consists of the trachea, bronchial tree, and lungs. Endowed with the function of mucociliary clearance, the nasal cavity, the trachea and bronchi constitute the first line of defense of the airway system toward external insults, its main role is to generate a sterile atmosphere. The lower respiratory tract represents the alveoli region, which is designed to perform gas exchange between the bloodstream and inhaled air.

The human respiratory tract model (HRTC) of the International Commission of Radiological Protection provides deposition data of inhaled particles from 1 nm to 10 μm of healthy adult female and male human subjects at different breathing patterns and physiological activities [4]. The HRTC clearly shows that particles deposit both in the upper and lower respiratory tracts with a proportion that varies in function of the size of the particles [5]. In an attempt to define the adverse effect of inhaled substances in vitro, ideally, an integrated testing strategy, which comprises both upper and low respiratory tract models, should be envisioned. This review describes the in vitro cell models currently available for inhalation toxicity purposes and will focus on the use of three-dimensional (3D) in vitro models of the upper respiratory tract made from primary human cells expressing a stable in vivo–like pattern of structures and functions. Several assays relevant for inhalation toxicity assessment will be discussed.

2.1.1 *Toolbox of in vitro Human Airway Models*

A survey of the existing models of in vitro inhalation toxicity testing may be useful for the readers. Knowing the strengths and limitations of each model will help scientists to choose the more appropriate tools for their work. There are several categories of human in vitro airway models:

- Cell lines
- Primary cells
- 3D cell models
- Cell cocultures
- Explants

In vivo, the cell composition of the surface of the respiratory tree plays an important role in maintaining the functionality of the lungs [6]. The alveoli consist mainly of alveolar type I and II cells. The alveolar epithelium is highly vascularized to ensure high gas exchange between air and blood. The epithelium, which covers the upper respiratory tract, is columnar and pseudostratified. The airway epithelium is composed of basal cells and ciliated and goblet cells producing mucus. Until now, the most popular human models are cell lines derived from various human tissues.

2.1.2 *Epithelial Cell Lines*

The airway epithelia constitute the first line of defense against external insults. It has a pseudo-layer structure consisting of three main types of cells: ciliated epithelial cells, mucus cells and basal cells. The mucus cells synthesize and secret mucin-rich mucus, which traps most of the inhaled particles, viruses, and bacteria; the latter are eliminated from the body by mucocilliary clearance by the cilia beating. All three cell types contribute to the pathogenesis of respiratory disease, for example, inflammatory reactions, mucus hypersecretion, and airway remodeling [7]. As airway epithelial cell lines, the most frequently used ones are A549, BEAS-2B and Calu-3. The characteristics and uses of these cell lines have been reviewed by Verstraelen et al. (2008) [8].

2.1.2.1 A549

Origin: This line was initiated in 1972 by Giard et al. [9] through explant culture of lung carcinomatous tissue from a 58-year-old Caucasian male. As type II pulmonary epithelial cells (alveolar pneumocyte), it synthesizes lecithin with a high percentage of unsaturated fatty acids (surfactants) and is oncogenic when tested in nude mice.

Applications: The cells can be used to screen chemical and biological agents for their ability to induce or affect differentiation and/or carcinogenesis—mechanistic studies, pathway mapping, and target finding or ranking of the toxicity potency of chemicals.

2.1.2.2 BEAS-2B

Origin: BEAS-2B cells were isolated from normal human bronchial epithelium obtained from the autopsies of noncancerous individuals. The cells were infected and immortalized with an adenovirus 12-SV40 virus hybrid Ad12SV40 and cloned. These cells retained the ability to undergo squamous differentiation in response to serum and stained positively for keratins and SV40 T antigen.

Applications: The cells can be used to test the toxicity of chemicals and biological agents relevant to upper-airway epithelia. BEAS-2B cells are suitable for mechanistic studies, pathway mapping and target finding.

2.1.2.3 Calu-3

Origin: Unlike most immortal cells, Calu-3 cells form sheets of cells that are welded to each other by tight junctions. These sheets form a fully functional epithelium that can transport large quantities of ions and fluid. In addition, Calu-3 cells have the highest level of natural cystic fibrosis transmembrane conductance regulator

(CFTR) expression of any known immortalized cell, even higher than some intestinal cell lines that once held the record. Thus, this cell line is suitable to study cystic fibrosis and drug discovery [10].

Applications: The cells can be used to test the drug transport across pulmonary epithelial cell monolayers [11].

2.1.3 *The Drawbacks of Cell Lines*

The drawbacks are as follows:

1. These cells have been transformed by oncogenes one way or other; thus certain signal transduction networks have been deregulated.
2. There are genetic aberrations such as chromosome loss, chromosome translation, mutations, etc.
3. These cell lines cannot give rise to fully differentiated phenotypes of the original tissue, such as cilia formation, mucus secretion, epithelium repair, and remodeling.
4. Under monolayer culture conditions, the cells behave totally differently compared to in vivo. These differences have been illustrated by comparing the response of the same cancer cells to drugs [12, 13].

2.1.4 *Immune Cells (Effector Cells)*

2.1.4.1 Dendritic cells

Although both the skin and the lung airway mucosa possess resident dendritic cells, the majority of studies conducted to date have utilized human peripheral blood mononuclear cell-derived dendritic cells (PBMC-DCs) due to their relative ease of extraction and ability to obtain larger quantities of cells [14]. Several protocols have been established to generate human DCs in vitro. Starting with blood or bone marrow–derived CD34+ hematopoietic progenitor cells (HPCs), DCs can be generated under various culture conditions using a cocktail of specific cytokines. Despite progress made in the field, it is still difficult to obtain sufficient amounts of primary DCs for basic or clinic research. Therefore, the use of cell lines such as THP-1, KG-1, and especially MUTZ-3 is invaluable [15].

2.1.4.2 Mast cells, neutrophils, eosinophils, and basophils

These are considered as effector cells, which are involved in early and late phases of asthma by releasing a plethora of inflammatory mediators. Their role in bronchoconstriction, mucus secretion, and airway remodeling is clearly defined. Many therapeutics target these effector cells as well as associated key molecules [16]. They can be isolated from the blood or cord blood and cultured in vitro. Their behavior such as migration, free radical production, viability, and apoptosis can be assessed after stimulation by allergens and cytokines [17].

2.1.5 *Coculture Models*

To study the cross talk between different types of cells, the simplest way is by coculturing different cells together in vitro. Macrophage activation plays an important role in the panel of pulmonary responses (being a major factor of phagocytosis) and is effective in initiating immune and inflammatory responses. Two cell-type coculture models, combining A549 cell lines with macrophages, have been developed to evaluate the cytoxicity of particles [18, 19]. Such experiments have been carried out using other established cell lines: for example, BEAS-2B were cocultured with human lung fibroblasts (HFL-1 or WISTAR-38) [20], human umbilical vein endothelial cells (ECV304) [21], or eosinophils [22], and primary human biliary epithelial cells (BECs) with alveolar macrophages [23].

Elegant triculture model systems including human blood–derived macrophages on top of A549 cells and monocyte-derived dendritic cells, have also been developed [24]. The potential of coculture in vitro models to study inflammatory and sensitizing effects of engineered nanoparticles or particulate matter has been expertly reviewed by Klein et al. [25].

2.1.6 *Precision-Cut Lung Slices*

The first use and characterization of precision-cut human lung slices in toxicology was reported in 1994 [26]. Slices are viable for at least three days. They can be placed under an inverted microscope, where digital image techniques allow quantification, not only of the responsiveness of single airways, but also the responsiveness of single vessels. In addition, it is possible to analyze the cilia-beating frequency. More than 20 slices can be obtained from one lung by this method. Tissue cores are prepared from the lungs filled with agarose solution after cooling to 4°C. From the cores, slices (220+/−20 μm) are cut [27]. Further development of the model allows one to study the responses of airways of different sizes (down to the terminal bronchioles) and to relate these changes in lung function to gene expression and mediator release. Natural innate cytokine responses to immunomodulators and adjuvants [28], drug metabolism studies [29], and early allergic response in small airways using precision-cut lung slices have also been reported [30].

2.1.7 *Lung-on-a-Chip*

Ingber's lab designed and constructed a microfluidics system, the so-called "Lung-on-a-Chip" that reproduces both the lung's alveolar–capillary interface and the mechanical effect of breathing on that interface—all on a polymer chip about 2 cm long [31]. Another example is a so-called multicompartmental bioreactor (MCB), designed on the basis of allometric scaling laws in order to recreate physiological life conditions of four different human cell types (pancreatic, adipocyte, endothelial, and hepatic cells), interconnected with each other through medium flow [32]. Some physiologically relevant results have been obtained using the MCB approach by Ahluwalia et al., suggesting that interactions, mediated by metabolites present in medium flow, have a remarkable effect—more than the physical interaction—and can lead to a restoration of physiological cell life conditions. By incorporating the immune cells into these microfluidics systems, it is possible to build an immune-competent human in vitro lung model for studying systemic toxicity.

2.1.8 *Fully Differentiated 3D Human Airway Epithelial Models*

To overcome the shortcomings of the cell lines, different techniques have been developed to make 3D cultures by providing a microenvironment or architecture closer to the in vivo situation. Among these techniques are cellular matrix scaffolds, hang-drop cultures, perfusion culture chambers and air–liquid interface (ALI) cultures.

However, to simulate in vivo lung conditions, ALI cultures seem to be more appropriate—the basal-lateral side of the epithelium is immersed in the culture medium, and the apical side is exposed to a humidified air/5% CO_2 environment. Furthermore, with Costar PET Transwell inserts, ALI cultures are very practical and convenient to use and suitable for most applications such as imaging, immunocytochemistry, toxicity tests, electrophysiological studies (using chamber measurement), and assessment of drug permeation and drug formulations.

Many in-house airway epithelial models based on primary human cells have been developed and used [33]. As an example, the following paragraphs focus on a commercially available and ready-to-use system named MucilAir (Epithelix, www.epithelix.com). MucilAir is a fully differentiated and ready-to-use 3D model of the human airway epithelium, constituted of primary human epithelial cells freshly isolated from nasal, tracheal or bronchial biopsies. MucilAir (Fig. 2.1) is not only morphologically and functionally differentiated but can also be maintained in a homeostatic state for a long period of time. Using cells from diseased donors, different versions of MucilAir can be made, such as asthmatic, allergic, chronic obstructive pulmonary disease (COPD), or cystic fibrosis models.

MucilAir is composed of basal cells, ciliated cells, and mucus cells. The proportion of these various cell types is preserved compared to what one observes in vivo. Moreover the epithelia are started from de-differentiated cells. The cells undergo progressive differentiation with time. After 45 days of culture the epithelia are fully ciliated and electrically tight (transepithelial electrical resistance [TEER] $\approx$ 450 $\Omega\,cm^2$), they secrete mucus and are ready to use. The epithelia are in a quasi-homeostatic state. They remain fully differentiated for more than one year, and one observes a harmonious balance between cell death and cell proliferation. However, MucilAir has a strong capability to regenerate after injuries (mechanical or chemical), even after elimination of half of the culture by scraping [34]. Electrophysiologically, the epithelia are fully functional: the activity of the main epithelial ionic channels, such as CFTR, EnaC, Na/K

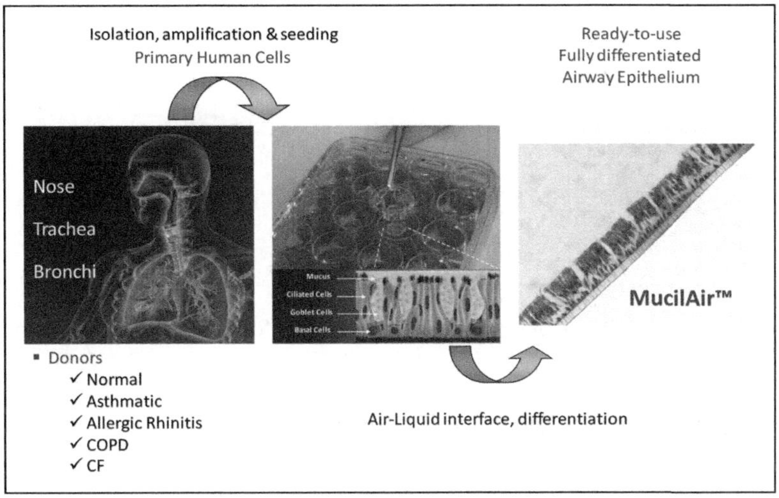

Figure 2.1 MucilAir, a fully differentiated 3D in vitro cell model of the human airway epithelia. Epithelial cells were freshly isolated from biopsies (nose, trachea, and bronchi) and then seeded onto a semiporous membrane (Costar Transwell, pore size 0.4 μm). After about 45 days of culture at an ALI, the epithelia were fully differentiated, both morphologically and functionally. Depending on the pathology of the donors, different versions of MucilAir could be made.

adenosine triphosphatase (ATPase), etc., is preserved. Moreover, the epithelia respond in a regulated and vectorial manner to the pro-inflammatory stimulus, tumor necrosis factor alpha (TNF-α). A large panel of cytokines, chemokines, and metalloproteinases has been detected in MucilAir (e.g., interleukin IL-8, IL-6; granulocyte-macrophage colony-stimulating factor GM-CSF; MMP-9, and GRO-α).

2.1.9 *Applications of MucilAir for Inhalation Toxicology*

Due to its fully differentiated nature, MucilAir can be used for studying various respiratory diseases. In the following paragraphs, we will give some examples about the applications of MucilAir in inhalation toxicology.

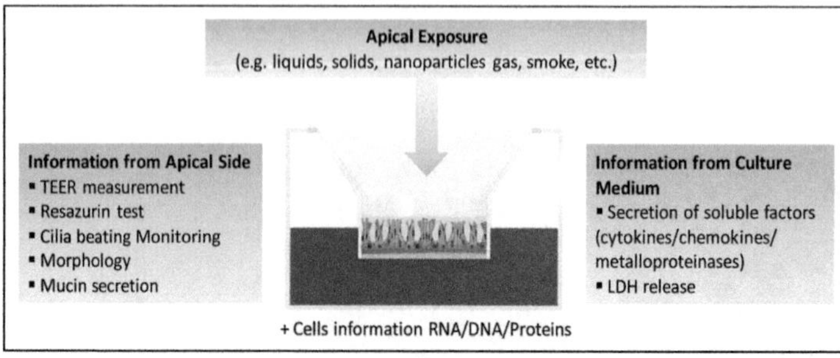

Figure 2.2 Multiple endpoint-testing strategy for toxicity testing using MucilAir.

2.1.10 *Multiple Endpoint Testing Strategy*

Fully differentiated 3D human airway epithelial models can also be used for testing the toxicity of drug candidates, formulations, or airborne xenobiotics. To this end, a multiendpoint strategy is used (Fig. 2.2). Xenobiotics can be applied on the apical surface as liquids, solids, gases, smoke, particulates, and nanoparticles. The toxicological effects can be monitored by several endpoints such as TEER, cell viability tests (Resazurin, lactate dehydrogenase [LDH]), cilia-beating frequency monitoring, mucociliary clearance, mucus secretion, and release of cytokines/chemokines.

One of the problems that in vitro toxicologists frequently encounter is the solubility of the tested compounds: not all chemicals are soluble or stable in solution. It is also difficult to test powders or nanoparticles. To solve this problem, a procedure that allows for the testing of all kinds of solid substances was developed (PCT/IB2010/053956). The idea is to use inert and neutral substances as carriers. Dextran was successfully used as a carrier to deliver the insoluble chemical compounds and nanoparticles onto the apical surfaces of MucilAir. Dextran is a bacterial by-product; the dextran macromolecule consists of glycan groups linked end to end. No harmful effect on MucilAir has been detected.

2.1.11 *Mucociliary Clearance and Cilia-Beating Frequency Analyses*

Mucociliary clearance is the principal defense mechanism of the respiratory system. Impairment of this function is the major cause of chronic infection and inflammation, as exemplified in cystic fibrosis disease. For experimental reasons, direct measurements of mucus velocity in human lungs are presently available only for the trachea, with values ranged from 4–6 mm/min [35]. The only reported value of the mucus velocity in the main bronchi is about 2.4 mm/min [36]. Interestingly, the same value has been obtained by using an in vitro culture model of the bronchial airway epithelial cells [37]. However, the velocity of mucociliary clearance varies considerably depending on individuals, anatomical location, and disease conditions. Indeed, the values obtained with MucilAir from the nasal epithelia are slightly inferior to that of bronchial ones, about 1.8 mm/min. Due to the simplicity of in vitro methods, it is possible to access the compounds effects on mucociliary clearance in various disease settings.

Two parameters predominantly drive the mucociliary clearance: (i) cilia-beating frequency and (ii) visco-elasticity of the mucus. As a single endpoint, modification of the cilia-beating frequency on the airway epithelia is added value information. Real-time measurement of cilia-beating activity on MucilAir can be performed using the method given in Ref. [38]. The effects of the chemical compounds can be assessed in real time in a noninvasive manner. The normal cilia-beating frequency found in MucilAir is between 7 Hz and 12 Hz.

2.1.12 *Respiratory Absorption*

The airway epithelium is a prime entry portal of xenobiotics into the body upon inhalation. Knowledge of toxicokinetics is needed to estimate the possible range of target doses at the cell or tissue level that can be expected from realistic external human exposure scenarios to inhaled compounds. This information is crucial for determining the dose range that should be used for in vitro testing. Kinetics in the in vitro system and dose response information is also crucial to translate in vitro results to the human in vivo situation [39].

By transposing the Caco-2 protocol [40] to MucilAir, transepithelial permeability of drug candidates or xenobiotics can be used to assess the upper-airway permeability of potential drugs, formulations, or xenobiotics. Measuring the rate of compound across MucilAir provides insight into the ability of different compounds to cross the respiratory epithelia. In permeability studies different compounds are tested at known concentrations and in two directions (apical to basolateral and basolateral to apical). To asset the rate of absorption, the permeability coefficient (P_{app}) is calculated for each compound. To assess whether a compound undergoes active efflux, apical (A) to basolateral (B) transport is measured (A→B) and vice versa (B→A). The ratio of B→A/A→B measurements (asymmetry index) is then determined. If the value of the asymmetry index is greater than 1.5, the compound is subject to active efflux.

2.1.13 *Detection of Airway Inflammation*

The airway epithelium is more than just a barrier; it is also a potent immunomodulator [41]. Upon external stimulation, it synthesizes and releases a large panel of chemokines, cytokines, lipids, growth factors, proteases, and protease inhibitors. These inflammatory mediators recruit the leukocytes to fight viruses and bacteria. This demonstrates an intact defense pathway typically for human airway epithelia. Applying a physiological saline solution (0.9% NaCl) on the apical surface of the epithelium results in a dramatic increase of IL-8 release. Indeed generating hypoxia conditions, the absence of air is sensed as a danger by the airway epithelia cells, defending themselves by the release of inflammatory mediators such as IL-8 [42]. In the epithelia, repetitive stimulation can be undertaken by cytomix, comprising TNF-α and lipopolysaccharide (LPS). After 24 h of stimulation with pro-inflammatory mediators, the amount of IL-8 released increased fivefold. Upon removal of the stimulus, the amount of IL-8 returned to basal levels as for day 0. The epithelia could respond to the stimulus in a physiological manner again and again. It is remarkable that the airway epithelial cells could recover after repetitive challenge of cytomix, a relatively physiological stimulus. This was not the case with other chemical compounds tested.

2.1.14 *Acute, Long-Term, and Repeated Dose Testing*

Following the testing strategy reported above, ranking the toxicity of chemicals [43] or drug candidates [44] in an acute setting has been

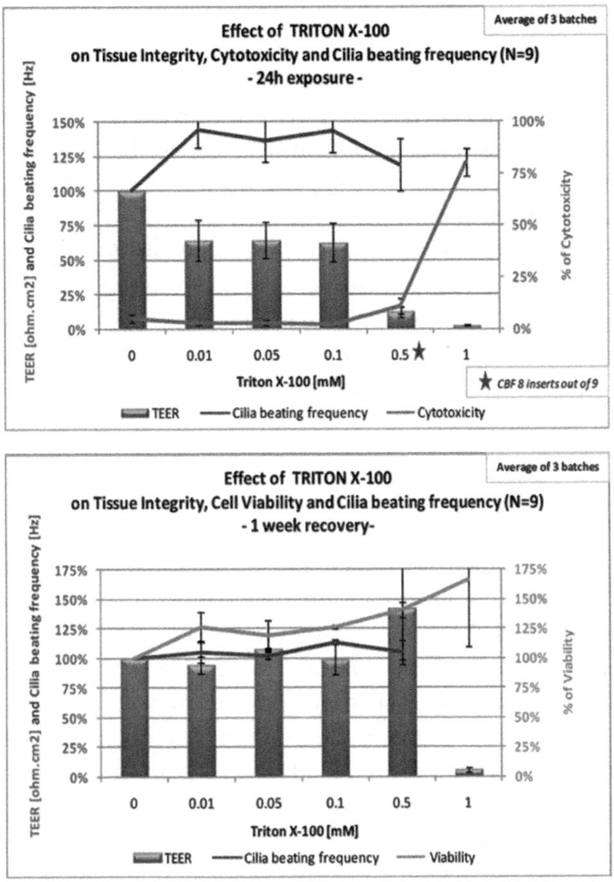

Figure 2.3 Example of reversibility of toxicity for concentrations around the EC50: after 24 h exposure of Triton X-100 at 0.5 mM, a loss of TEER as well as a slight increase of cytotoxicity were observed. After one week recovery, the epithelium recovered showing a TEER above 10%.

reported. Chemicals are exposed at the apical surface of MucilAir for 1 or 24 h and then washed to mimic natural mucociliary clearance, the endpoints being measured 24 h after the exposure. Interestingly, reversibility of parameters like TEER, cilia-beating frequency, or cytokine release can be evaluated (Fig. 2.3).

Quite often, it is necessary to perform repeated dose and long-term toxicity/efficacy tests of drug candidates. Up to now, these kinds of experiments can be performed only on animals. In the following graph, we provide a proof-of-concept for repeated dose tests using an in vitro cell model. It is a transposition of the OECD413 guidelines for a subchronic 90-day test on rodents. TEER was used as an endpoint, which is an indicator of the epithelial integrity. This is a very sensitive endpoint. Repeated dose tests of formaldehyde on MucilAir were at 90 days (repeated dose exposure study). The 6 h/day exposure on MucilAir to formaldehyde for a period of 90 days is shown in Fig. 2.4. Every day, tissue integrity (TEER) was measured ($n = 3$), and then epithelia were reused for the next exposure. The "no observed adverse effect level" (NOAEL) can also be determined (around 1 mM).

2.2 Conclusions

Each model has its strengths and weaknesses, and there is no perfect respiratory model for toxicity assessment. Depending on the goal or on application, one model could be better than another. Since airborne pollutants, particles, fumes, gases, etc., affect human beings, it is logical to assess the toxicity and efficacy on cells of human origin, especially in 3D human models. Progress has to be made by scientists to build integrated testing strategies based on complementary airway models. Particular efforts have to be made toward the development of in vitro tests to assess respiratory irritation and sensitization. Due to its central role in pathogenesis of respiratory disease, the 3D in vitro cell models of the human airway epithelium deserve more attention in the future. We believe that 3D cell models like MucilAir are highly relevant and valuable for the development of new drugs, as well as assessing the effect of airborne xenobiotics.

Acknowledgments

The authors thank Ligue Suisse Contre la Vivisection and Fondation E. Naëf pour la recherche in vitro (Geneva) for their continuous support.

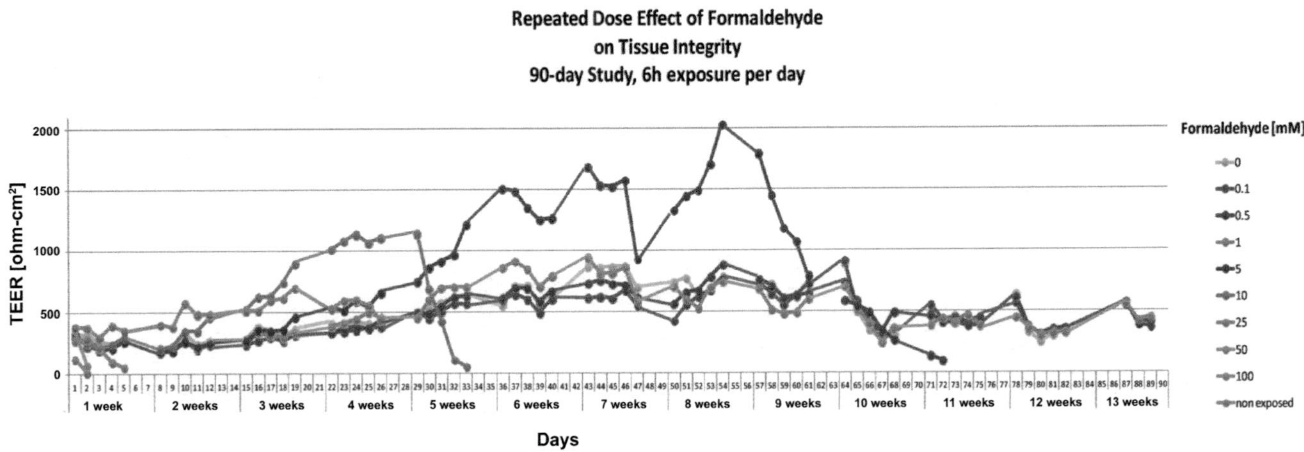

Figure 2.4 Repeated dose tests of formaldehyde on MucilAir (90 days' repeated dose exposure study). The 6 h/day exposure on MucilAir to formaldehyde for a period of 90 days. Every day, tissue integrity (TEER) was measured ($n = 3$) and epithelia were then reused for the next exposure.

References

1. Committee on Toxicity Testing and Assessment of Environmental Agents, Board on Environmental Studies and Toxicology, Institute for Laboratory Animal Research, National Research Council (2007) *Toxicity Testing in the 21st Century. A Vision and a Strategy*, Washington, DC: National Academies Press.
2. BéruBé, K., Aufderheide, M., Breheny, D., Clothier, R., Combes, R., Duffin, R., Forbes, B., Gaça, M., Gray, A., Hall, I., Kelly, M., Lethem, M., Liebsch, M., Merolla, L., Morin, J. P., Seagrave, J., Swartz, M. A., Tetley, T. D., Umachandran, M. (2009) In vitro models of inhalation toxicity and disease. The report of a FRAME workshop. *Altern Lab Anim* **37**, 89–141.
3. Roggen, E. L. (2011) In vitro toxicity testing in the twenty-first century. *Front Pharmacol* **2**, 3.
4. International Commission on Radiological Protection (1994) Human respiratory tract model for radiological protection. CRP Publication 66. A report of a task group of the ICRP. *Ann ICRP* **24**, 1–482.
5. Geiser, M., Kreyling, W. G. (2010) Deposition and biokinetics of inhaled nanoparticles. *Part Fibre Toxicol* **7**, 2.
6. Ochs, M., Webel, E. R. (2008) Functional design of the human lung for gas exchange. In Fishman, A. P. (ed.), *Fishman's Pulmonary Diseases and Disorders*, McGray-Hill Medical.
7. Epstein, M. M. (2004) Do mouse models of allergic asthma mimic clinical disease? *Int Arch Allergy Immunol* **133**, 84–100.
8. Verstraelen, S., Bloemen, K., Nelissen, I., Witters, H., Schoeters, G., Van Den Heuvel, R. (2008) Cell types involved in allergic asthma and their use in in vitro models to assess respiratory sensitization. *Toxicol in Vitro* **22**, 1419–1431.
9. Giard, D. J., Aaronson, S. A., Todaro, G. J., Arnstein, P., Kersey, J. H., Dosik, H., Parks, W. P. (1973) In vitro cultivation of human tumors: establishment of cell lines derived from a series of solid tumors. *J Natl Cancer Inst* **51**, 1417–1423.
10. Haws, C., Finkbeiner, W. E., Widdicombe, J. H., Wine, J. J. (1994) Calu-3: a human airway epithelial cell line that shows cAMP-dependent Cl-secretion. *Am J Physiol* **266**, 493–501.
11. Bur, M., Huwer, H., Muys, L., Lehr, C. M. (2010) Drug transport across pulmonary epithelial cell monolayers: effects of particle size, apical liquid volume, and deposition technique. *J Aerosol Med Pulm Drug Deliv* **23**, 119–127.

12. Bissell, M. J., LaBarge, M. A. (2005) Context, tissue plasticity, and cancer: are tumor stem cells also regulated by the microenvironment? *Cancer Cell* **7**, 17–23.

13. Griffith, L. G., Swartz, M. A. (2006) Capturing complex 3D tissue physiology in vitro. *Nat Rev Mol Cell Biol* **7**, 211–224.

14. Casati, S., Aeby, P., Basketter, D. A., Cavani, A., Gennari, A., Gerberick, G. F., Griem, P., Hartung, T., Kimber, I., Lepoittevin, J. P., Meade, B. J., Pallardy, M., Rougier, N., Rousset, F., Rubinstenn, G., Sallusto, F., Verheyen, G. R., Zuang, V. (2005) Dendritic cells as a tool for the predictive identification of skin sensitisation hazard Report and recommendations of ECVAM Workshop 51. *Altern Lab Anim* **33**, 47–62.

15. Santegoets, S. J., van den Eertwegh, A. J., van de Loosdrecht, A. A., Scheper, R. J., de Gruijl, T. D. (2008) Human dendritic cell line models for DC differentiation and clinical DC vaccination studies. *J Leukoc Biol* **84**, 1364–1373.

16. Casale, T. B., Stokes, J. R. (2008) Immunomodulators for allergic respiratory disorders. *J Allergy Clin Immunol* **121**, 288–296.

17. Nilsson, C., Larsson, A. K., Höglind, A., Gabrielsson, S., Troye Blomberg, M., Lilja, G. (2004) Low numbers of interleukin-12-producing cord blood mononuclear cells and immunoglobulin E sensitization in early childhood. *Clin Exp Allergy* **34**, 373–380.

18. Wottrich, R., Diabaté, S., Krug, H. F. (2004) Biological effects of ultrafine model particles in human macrophages and epithelial cells in mono- and co-culture. *Int J Hyg Environ Health* **207**, 353–361.

19. Crabbé, A., Sarker, S. F., Van Houdt, R., Ott, C. M., Leys, N., Cornelis, P., Nickerson, C. A. (2011) Alveolar epithelium protects macrophages from quorum sensing-induced cytotoxicity in a three-dimensional co-culture model. *Cell Microbiol* **13**, 469–481.

20. Lang, D. S., Jörres, R. A., Mücke, M., Siegfried, W., Magnussen, H. (1998) Interactions between human bronchoepithelial cells and lung fibroblasts after ozone exposure in vitro. *Toxicol Lett* **96–97**, 13–24.

21. Mögel, M., Krüger, E., Krug, H. F., Seidel, A. (1998) A new coculture-system of bronchial epithelial and endothelial cells as a model for studying ozone effects on airway tissue. *Toxicol Lett* **96–97**, 25–32.

22. Wong, C. K., Wang, C. B., Ip, W. K., Tian, Y. P., Lam, C. W. (2005) Role of p38 MAPK and NF-kB for chemokine release in coculture of human eosinophils and bronchial epithelial cells. *Clin Exp Immunol* **139**, 90–100.

23. Ishii, H., Hayashi, S., Hogg, J. C., Fujii, T., Goto, Y., Sakamoto, N., Mukae, H., Vincent, R., van Eeden, S. F. (2005) Alveolar macrophage-epithelial cell interaction following exposure to atmospheric particles induces the release of mediators involved in monocyte mobilization and recruitment. *Respir Res* **6**, 87.
24. Rothen-Rutishauser, B. M., Kiama, S. G., Gehr, P. (2005) A three-dimensional cellular model of the human respiratory tract to study the interaction with particles. *Am J Respir Cell Mol Biol* **32**, 281–289.
25. Klein, S. G., Hennen, J., Serchi, T., Blömeke, B., Gutleb, A. C. (2011) Potential of coculture in vitro models to study inflammatory and sensitizing effects of particles on the lung. *J Immunol* **186**, 2238–2244.
26. Fisher, R. L., Smith, M. S., Hasal, S. J., Hasal, K. S., Gandolfi, A. J., Brendel, K. (1994) The use of human lung slices in toxicology. *Hum Exp Toxicol* **13**, 466–471.
27. Martin, C., Uhlig, S., Ullrich, V. (1996) Videomicroscopy of methacholine-induced contraction of individual airways in precision-cut lung slices. *Eur Respir J* **9,** 2479–2487.
28. Switalla, S., Lauenstein, L., Prenzler, F., Knothe, S., Förster, C., Fieguth, H. G., Pfennig, O., Schaumann, F., Martin, C., Guzman, C. A., Ebensen, T., Müller, M., Hohlfeld, J. M., Krug, N., Braun, A., Sewald, K., (2010) Natural innate cytokine response to immunomodulators and adjuvants in human precision-cut lung slices. *Toxicol Appl Pharmacol* **246**(3), 107–115.
29. De Kanter, R., Olinga, P., De Jager, M. H., Merema, M. T., Meijer, D. K., Groothius, G. M. (1999) Organ slices as an in vitro test system for drug metabolism in human liver, lung and kidney. *Toxicol in Vitro* **13**, 737–744.
30. Wohlsen, A., Martin, C., Vollmer, E., Branscheid, D., Magnussen, H., Becker, W. M., Lepp, U., Uhlig, S. (2003) The early allergic response in small airways of human precision-cut lung slices. *Eur Respir J* **21**, 1024–1032.
31. Huh, D., Matthews, B. D., Mammoto, A., Montoya-Zavala, M., Yuan Hsin, H., Ingber, D. E. (2010) Reconstituting organ-level lung functions on a chip. *Science* **328**, 1662–1668.
32. Vozzi, F., Heinrich, J. M., Bader, A., Ahluwalia, A. D. (2009) Connected culture of murine hepatocytes and HUVEC in a multi-compartmental bioreactor. *Tissue Eng Part A* **15**, 1291–1299.
33. Bérubé, K., Prytherch, Z., Job, C., Hughes, T. (2010) Human primary bronchial lung cell constructs: the new respiratory models. *Toxicology* **278**, 311–318.

34. Crespin, S., Bacchetta, M., Huang, S., Dudez, T., Wiszniewski, L., Chanson, M. (2011) Approaches to study differentiation and repair of human airway epithelial cells. *Methods Mol Biol* **742**, 173–185.

35. Hofmann, W., Asgharian, B., Winkler-Heil, R. (2002). Inter-subject variability in particle deposition in human lungs. *J Aerosol Sci* **33**, 219 –235.

36. Foster, W. M., Langenback, E. G., Bergofsky, E. H. (1982) Lung mucociliary function in man: interdependence of bronchial and tracheal mucus transport velocities with lung clearance in bronchial asthma and healthy subjects. *Ann Occup Hyg* **26**, 227–244.

37. Matsui, H., Grubb, B. R., Tarran, R., Randell, S. H., Gatzy, J. T. (1998) Evidence for periciliary liquid layer depletion, no abnormal ion composition, in the pathogenesis of cystic fibrosis airways disease. *Cell* **95**, 1005–1015.

38. Dimova, S., Maes, F., Brewster, M. E., Jorissen, M., Noppe, M., Augustijns, P. (2005) High-speed digital imaging method for ciliary beat frequency measurement. *J Pharm Pharmacol* **57**(4), 521–526.

39. Adler S. et al. (2011) Alternative (non-animal) methods for cosmetics testing: current status and future prospects—2010. *Arch Toxicol* **85**, 367–485.

40. Hubatsch, I., Ragnarsson, E. G. E., Artursson, P. (2007) Determination of drug permeability and prediction of drug absorption in Caco-2 monolayers. *Nature Prot* **2**, 2111–2119.

41. Huang, S., Wiszniewski, L., Derouette, J. P., Constant, S. (2009) In vitro organ culture models of asthma. *Drug Disc Today: Dis Models* **6**, 137–144.

42. Huang, S., CaulFuty, M. (2009) A novel in vitro cell model of the human airway epithelium. *3R-Info-Bulletin* **41**(October 2009).

43. Constant, S., Huang, S., Derouette, J. P., Wiszniewski, L. (2008) MucilAir: a novel in vitro human 3D airway epithelium model for assessing the potential hazard of nanoparticles and chemical compounds. *Toxicol Lett* **180**, S233.

44. Maier, M. S. V. (2011) Setting occupational exposure limits for unstudied pharmaceutical intermediates using an in vitro parallelogram approach. *Toxicol Mech Methods* **21**, 76–85.

Chapter 3

Cultivation of Human Hepatocytes in the Quasi-Vivo® System: From Isolation and Seeding to Quantification of Xenobiotic-Metabolizing Enzyme Expression and Activity

Cedric Duret,[a] Bruna Vinci,[b] Tommaso Sbrana,[c] Sylvie Klieber,[d] Patrick Maurel,[a] Martine Daujat-Chavanieu,[a] and Arti Ahluwalia[c]

[a]*Inserm, U1040, Montpellier, France, and CHU Montpellier, Institut de Recherche en Biothérapie, Hôpital Saint Eloi, Montpellier, France*
[b]*CNR Institute of Clinical Physiology, Pisa, Italy*
[c]*Interdepartmental Research Center "E.Piaggio", Faculty of Engineering, University of Pisa, Via Diotisalvi, 2 - 56126 Pisa, Italy*
[d]*Sanofi, Recherche et Dévelopement, Département DMPK-S, Montpellier, France*
patrick.maurel@inserm.fr

In vitro liver models for toxicity testing suffer from a number of drawbacks, including short-term viability and phenotypic changes mainly associated with huge drops in P450 expression of hepatocytes [1]. The powerful but sensitive metabolic machinery of hepatocytes depends on the orchestration of a multitude of signals and stimuli present in vivo. The in vitro microenvironment differs greatly from the physiological ambient due to the lack of

Cellular in vitro Testing: Methods and Protocols
Edited by John W. Haycock, Arti Ahluwalia, and John M. Wilkinson

ISBN 978-981-4364-97-3 (Hardcover), 978-981-4364-98-0 (eBook)
www.panstanford.com

three-dimensional and dynamic cues, as well as the absence of other cell types. With its nonturbulent, high flow rate and low physiological shear stress [2], the Quasi-Vivo® system is the ideal environment for maintaining the expression of drug-metabolizing enzymes in human hepatocytes, which are considered the closest in vitro model of the human liver [3]. Here we describe how human adult hepatocytes can be isolated from liver resections and cultured in pseudo three-dimensional collagen sandwiches under dynamic flow conditions for long periods. Methods for quantifying the expression of xenobiotic-metabolizing enzymes (XMEs) and drug metabolism in Quasi-Vivo® are also presented. The methods can be employed for 3Rs-compliant preclinical drug testing since the use of animals in the drug-testing process can be significantly reduced.

3.1 Human Liver Isolation

Adult human liver tissue suitable for the isolation of hepatocytes is provided from donors undergoing surgical liver resection in general for the removal of metastatic tumors or liver diseases (e.g., hydatid cyst). Use of human liver samples for hepatocyte preparation for scientific purposes must be approved by the appropriate national and local ethics committees and by other local regulatory authorities. No information on the patients must be available in the laboratory, apart from sex, age, and the reason for surgical resection. The patients cannot be identified, directly or through identifiers. Written consent of patients is necessary.

During hepatic resection, the part of the tissue encompassing the tumor(s) is dissected by the surgeon in the operation theater and sent to the anatomopathologists, while the remaining downstream encapsulated part is used for hepatocyte preparation. The quality of the perfusion is critical for hepatocyte isolation for this reason; pieces of liver ranging from 50 g to 300 g should be preferred. Samples must be encapsulated in Glisson's capsule on all areas except, obviously, the edge left by the surgeon. There must be several veins apparent on the cut edge; these will be used for perfusion. When the tissue is collected it is placed in a sterile plastic bag on ice, in the presence of a sufficient liquid volume of physiologic serum (9 g/L NaCl in sterile water) to overlay it. The sample is generally used within one to 8 h of removal (time for transport from the hospital to the laboratory).

3.2 Materials

3.2.1 *Equipment for Liver Isolation*

The hepatocyte isolation procedure must be performed in a cell culture room (safety grade level 2 in France). All materials used are listed below:

1. Standard equipment for the cell culture room: A class II biosafety cabinet, a CO_2 incubator, a phase-contrast optical microscope, and a refrigerated centrifuge with rotor, buckets, and adaptors to accommodate 50–250 mL centrifuge tubes for cell sedimentation.
2. Protective gear and classical laboratory devices: Safety glasses, laboratory coveralls, shoe covers, bonnets, surgical masks, protective sleeves, disposable pipettes, etc.
2. Suitable surgical instruments (scalpels, forceps, sterile chirurgical gloves).
4. A suitable apparatus to include a platform for the liver undergoing perfusion and digestion, peristaltic pumps capable of flow rates of 10–500 mL/min, and a heating unit to maintain the temperature of the system at a constant 36–38°C.
5. A water bath for maintaining perfusion buffers at 37°C.
6. A pump for the liquid aspiration device (for removal of liquid waste).
7. Stainless steel bowls and a 250 μm strainer that can be decontaminated and sterilized by autoclaving to reuse.
8. Masterflex biocompatible tubing (ID 4.8 mm, AD 8.0 mm) and tubing adaptors that can enter vessels.
9. Cannulas for aspiration.
10. Multiwell plates (Biocoat®, BD Biosciences).

3.2.2 *Reagents and Buffer for Cell Isolation*

Buffers and solutions are prepared with deionized water, sterilized by passing through 0.22 μm filters.

1. *N* - (2 - hydroxyethyl)piperazine - *N*′ - (2 - ethanesulfonic acid) (HEPES) buffer: 10 mM HEPES (Sigma-Aldrich), 136 mM NaCl (Sigma-Aldrich), 5 mM KCl (Sigma-Aldrich), and 0.5% glucose (Sigma-Aldrich), pH 7.6.
2. Ethylenebis(oxyethylenenitrilo) tetra-acetic acid (EGTA) buffer: 0.5 mM EGTA (Sigma-Aldrich) in HEPES buffer (described in #1 of this section).
3. Collagenase buffer: HEPES buffer (described in #1 of this section), with collagenase type IV (Sigma-Aldrich) at an activity concentration of 2×10^5 IU/L.
4. Wash solution: HEPES buffer (described in #1 of this section) complemented with 2.5% fetal bovine serum (FBS; Invitrogen). Add FBS just before liver perfusion.
5. Aliquots (10 mL) of $CaCl_2$ solution at 70 mM. Dissolve in deionized water and filter on a 0.22 μm filter syringe. Just before liver perfusion, add 10 mL of 70 mM $CaCl_2$ in 1 L of collagenase buffer (#3 of this section).
6. Antibiotic solution: Penicillin/streptomycin (Invitrogen). Add 10 mL/L to HEPES buffer, EGTA buffer, and collagenase buffer and wash the solution just before liver perfusion.
7. Antifungic solution: Fungizone (Invitrogen). Add 3 mL/L of 250 μg/mL Fungizone to HEPES buffer, EGTA buffer, and collagenase buffer and wash the solution just before liver perfusion.
8. Trypan Blue solution 0.4% (Invitrogen) filtered on a 0.45 μm filter syringe.
9. Plating medium: Ham's F12/William's E medium (1:1) supplemented with 2.5% FBS (v/v), bovin serum albumin 15 μg/mL, 66.5 μM ethanolamine, 50 mg/L transferrin, 7.2 μM linoleic acid, 2 mg/L insulin, 0.2 mg/mL glucagon, 0.1 μM dexamethasone, 1.26 g/L glucose, 0.4 mM sodium pyruvate, and 50 mg/L ascorbic acid (all from Sigma).

3.3 Methods

3.3.1 *Liver Perfusion and Hepatocyte Enrichment*

1. Place all buffers except the wash solution (point 4, Section 3.2.2) at 37°C at least 45 min before liver perfusion. The wash solution is used at room temperature.

2. Set up the perfusion apparatus inside the biosafety cabinet, including the heating apparatus, peristaltic pump, and tubing with adaptors and cannula.
3. Transfer the liver sample to the perfusion vessel and examine the edge carefully to locate the various vein and artery entries that will be used for perfusion.
4. First wash and warm the liver by perfusion of 1 L HEPES buffer, flow rate 50–100 mL/min, with no recirculation. Insert the adaptor at the end of the tubing successively in all veins/arteries present on the edge.
5. Second, perfuse the sample with 1 L of EGTA buffer, flow rate 50–100 mL/min, with no recirculation. Insert the adaptor at the end of the tubing successively in all veins/arteries present on the edge.
6. Third, wash the tissue (remove the EGTA compound) by perfusion of 1 L HEPES buffer, flow rate 50–100 mL/min, with no recirculation. Insert the adaptor at the end of the tubing successively in all veins/arteries present on the edge. At the end of this step, empty the reservoir of the perfusion vessel is wash it several times with sterile water.
7. Fourth, digest the liver by perfusion with collagenase buffer, flow rate 50–100 mL/min, with recirculation. Insert the adaptor at the end of the tubing successively in all veins/arteries present on the edge. The duration of this step lasts for a maximum of 20 min.
8. Transfer the liver to a stainless steel bowl, break Glisson's capsule manually, and gently disrupt the tissue with the help of manipulator hands.
9. Complement the homogenate with 500 mL of wash buffer and filtrate it on a 250 μm stainless steel strainer.
10. Wash the filter twice with approximately 200 mL of wash buffer and distribute the filtrate into 250 mL centrifuge tubes.
11. Centrifuge the tubes for 5 min at 50 g at 10–15°C.
12. The pellets contain hepatocytes. Discard the supernatant and resuspend the pellets in wash buffer. Repeat steps 11 and 12 two times.
13. After the third centrifuge, resuspend pellets in a plating medium and pool all in one conical tube.

14. Perform a cell count and viability assessment by trypan blue exclusion using a hemocytometer.
15. Determine the total cell yield, percent viability, and viable cell concentration.

3.3.2 *Long-Term Medium*

For hepatocyte culture in the Quasi-Vivo® system we used long-term media described in Pichard et al. [4].

Long-term medium: William's E medium, 1.25 μM ethanolamine, 5 mg/L transferrin, 15.4 μM linoleic acid, 10 mg/L insulin, 0.1 μM selenium acid, 1 mg/mL glucagon, 0.1 μM dexamethasone, cAMP 5 μM, prolactin 10 IU/L, 2 g/L glucose, epidermal growth factor (EGF) 50 μg/l, liver growth factor (LGF) 20 μg/L, 0.25 mM sodium pyruvate, and 2 mg/L ascorbic acid (all from Sigma)

3.3.3 *Quasi-Vivo® Experiments*

The Quasi-Vivo® chambers, illustrated in Fig. 3.1, are the basic building blocks of the system. They can be connected in series with up to 16 chambers. As a rule of thumb, each chamber requires

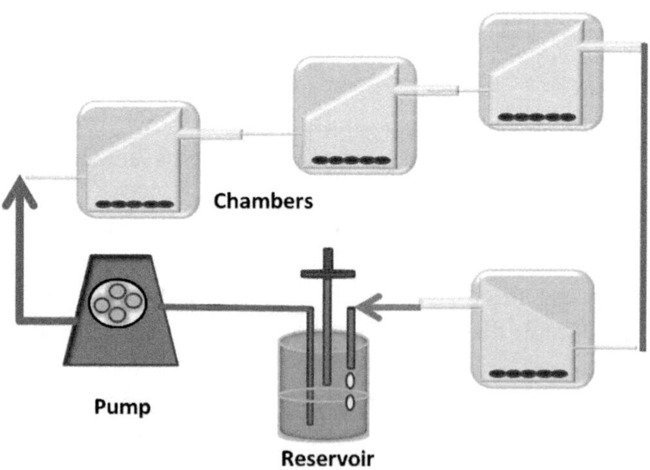

Figure 3.1 Schematic of a four-chamber Quasi-Vivo® circuit and connection scheme.

2 mL of medium; 1 mL is necessary for the tubing, while the reservoir bottle requires 2 mL. If long-term experiments (over 48 h) are conducted it is necessary to add more medium in the case of evaporation. The best method for sterilizing the system is using gas plasma. If this method is not available, the complete circuit should be sterilized using an autoclave. This procedure can be repeated up to 10 times.

3.3.4 *Controls*

1. Appropriate controls are essential for demonstrating that the Quasi-Vivo® system is more effective than static cell cultures. One can choose to assess controls using a setup in which the control volume is equal to that of the Quasi-Vivo® system or controls using multiwells, in which the total volume is less. In the latter case, it should be noted that protein or metabolite secretion by cells into the medium will be more dilute and that the medium may need to be changed more often.
2. Therefore equal volume static control, as described below, is preferred. In equal volume "static control" the volume of the culture medium is the same as that used in the Quasi-Vivo® circuit. It is then important to maintain this volume equal in both experiments. Figure 3.2 illustrates this concept.
3. During the experiment, as chambers are removed from the circuit for ribonucleic acid (RNA) quantification, the same volume of medium must be removed from the control. This procedure is necessary to maintain the allostatic balance in the control, which enables comparisons between the dynamic Quasi-Vivo® system and the same cells from the same donor, the same number of cells, the same timing, and the same coating and sandwich with the same media volume but in the absence of flow conditions.
4. Conversely, the "volume control" is represented by an experiment conducted in a multiwell at the same time as the "dynamic" experiment in the Quasi-Vivo® system (cells from the same donor, the same number of cells, the same timing, and the same coating and sandwich) with a volume of 1 mL per slide (typical volume of culture medium for classical culture in static conditions). This procedure allows the evaluation of the changes

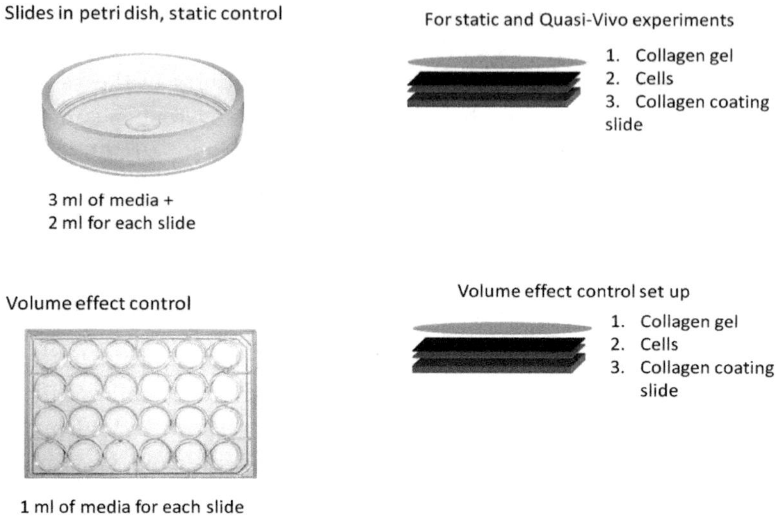

Figure 3.2 Schematic of two alternative control setups.

in cell behavior under conditions of greater volume, which is necessary for culture in the Quasi-Vivo® setup.

5. Note that two variables are changed in the Quasi-Vivo® system with respect to this control, flow and volume.

3.3.5 *Preparation of Glass Slides for Seeding*

Glass or plastic slides are seeded with hepatocytes in a collagen sandwich and placed in the chambers.

1. Use 12 mm diameter glass slides (Marienfeld GmbH) or sterile, pretreated plastic slides (Nunc).
2. For slide coating with rat collagen type I and cell seeding, place one slide in each well of a 24-microwell plate.
3. Note, if glass slides are used a sterilization step is required:
    a. Rinse the slides two to three times with phosphate buffered saline (PBS) and leave them to dry under a hood.
    b. Sterilize the slides using an autoclave (135°C at 2 atm for 20 min or preferably using gas plasma since it improves the adhesive properties of the surface).

4. From this moment on, ensure that sterile conditions are maintained.

3.3.5.1 Coating of slides

1. Coat the upper surface of the slides with 200 μL of type I collagen (Becton Dickinson), 50 μg/mL in sterile acetic acid 0.02 N.
2. Incubate the slides at 37°C with the collagen for at least 1 h.
3. Under the hood, aspirate the solution of collagen, rinsing four to five times with 0.5 mL of sterile PBS. The last aliquot of PBS is left in contact with the slides until actual use of the support.
4. Just before use, aspirate PBS and leave the slides to dry under a sterile hood.

3.3.6 *Hepatocyte Plating and Culture*

1. After evaluation of yield and viability (Section 3.3.1, #14–15), hepatocytes are diluted in a plating medium at a concentration of 1×10^6 viable cells/mL. Then 0.5 mL of a hepatocyte-diluted suspension is added per well of a 24-well plate with a collagen coat (Biocoat, BectonDickinson) or a 24-well plate containing a coverslip collagen coat. This number of cells per well corresponds approximately to a cell density of 12.5×10^4 cells per cm^2 for a confluent monolayer. Care must be taken to rehomogenize the cell suspension frequently by gentle agitation (inversion) whilst distributing to the culture multiwell.
2. Cells are evenly distributed on the well by gentle agitation. The hepatic differentiated phenotype is maintained in confluent but not in subconfluent cultures. Establishment of a homogeneous confluent-cell monolayer throughout the dish is therefore critical. Once the slides present on one tray are supplemented with the culture medium and cells, the tray is held horizontally and gently agitated from left to right (1 cycle/s for 10 s) and back and forth (1 cycle/s for 10 s) with a pause of 5 seconds in between. This allows the suspension to be spread homogeneously on the dish and avoids rotational movement of the suspension, which would

result in preferential distribution of cells toward the periphery with low density in the center.

3. Culture dishes are then placed in an incubator in a humid atmosphere of air 5% CO_2 at 37°C.
4. After overnight seeding time, the serum-supplemented plating medium is discarded and replaced by a long-term medium. Note: The culture medium must be aspirated and poured gently to avoid detachment of cells. Medium changes will be facilitated by placing the trays on an inclined plane.

3.3.7 *Quasi-Vivo® Setup for Hepatocyte Culture*

1. The Quasi-Vivo® system is shown in Fig. 3.1. Note that the Quasi-Vivo® chambers are not symmetrical. The input tube is smaller in diameter than the output tube. Therefore it is important when connecting chambers in series in order to ensure that the output of one chamber is connected to the input of the following chamber. Figure 3.1a indicates how the system should be assembled, and Fig. 3.3 shows a typical 4-chamber circuit.

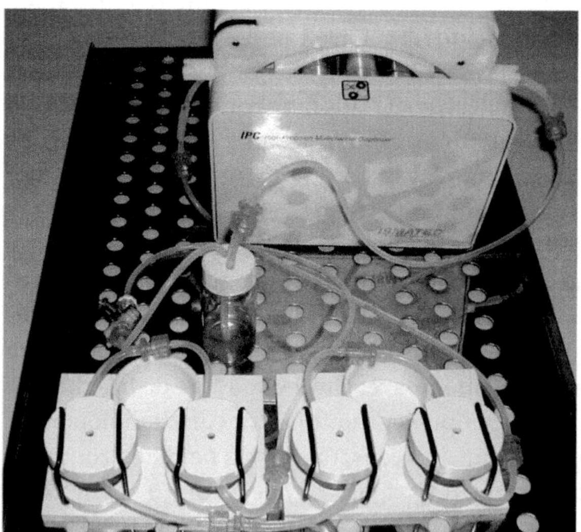

Figure 3.3 An assembled Quasi-Vivo® circuit with four chambers in series.

2. Prior to running experiments in the Quasi-Vivo® system it is essential to set the pump parameters (revolutions per minute [rpm]) in order to generate the desired flow rate in the circuit. The pump rpm will also depend on the number of chambers in the system and the length of tubing. We recommend the Ismatech IPC4 because it is small enough to sit beside the Quasi-Vivo® circuit in a standard-size incubator.

3.3.8 *Collagen Sandwich*

Sandwich collagen cultures of human primary hepatocytes are necessary to avoid cell detachment by the action of media flow. Furthermore it is well known that hepatic phenotypic functions are better conserved when hepatocytes are placed between two layers of collagen gel, since this more closely mimics the in vivo environment [5].

1. Prepare a cold (on ice, 4°C) solution containing the following amounts of reagent per coverslip:

   a. Rat collagen type I (Becton Dickinson) 250 μg
   b. PBS 10X 20 μL
   c. NaOH 1 N 1.53 μL

   Bring the volume to 300 μL with sterile distilled water. Prepare a global sterile mix for all coverslips and keep it cold until use (less than 2 h after preparation).
2. Just before adding the collagen cover, carefully transfer the slides to the Quasi-Vivo® chambers (Fig. 3.4) or to the static control petri dish.
3. Distribute 200 μL per well (24-well plate) or 200 μL per hepatocyte-containing coverslip.
4. Incubate the slides at 37°C with the collagen for at least 30 min.
5. Under the hood, aspirate the solution of collagen and add 1 mL of a long-term medium per well. Add the same volume to the static control dish as in the Quasi-Vivo® circuit.
6. Renew the multiwell, static control dish and the Quasi-Vivo® system culture medium every week.

Figure 3.4 The slide is carefully inserted into the bottom of the chamber.

3.3.9 *Flow Circuit Preparation*

1. Place the chambers on the base ring of the clamping tray and place the reservoir bottle in its holder.
2. Open the chambers and put some drops of the medium onto the bottom to prevent drying out of the cells surface while filling the circuit. Carefully open the bottle in a sterile way and add the rest of the medium.
3. Close the chambers with clamps and rubber rings. Connect the chambers in series, paying attention to the direction of the flow, as indicated in Fig. 3.1.
4. Turn on the pump at a flow rate of 100–250 μL/min. Once small drops of the medium are observed to come out of the short tube, the circuit is full. Ensure that the longer tube of the reservoir is always in the medium and the shorter one above the level of the medium in the bottle. As long as droplets continue to exit from the short tube into the bottle, and there are no leaks, the system is running correctly.
5. Put the system on the pump and the multiwells with controls on a removable tray in the incubator for the experimental time (up to 21 days, removing 2 chambers at a time, every 2–3 days for a quantitative polymerase chain reaction [qPCR]).

3.3.10 *Removal of Chambers for qPCR*

1. Remove the system from the incubator and switch off the pump. Place the tray under a laminar flow hood. Carefully disconnect the

two chambers by opening the clamping ring and then the upper side of the chambers without contaminating the circuit or losing the medium.

2. Being careful to remove only 4 mL of the medium for sampling, immediately reconnect the remaining chambers. The medium can be removed by attaching a syringe to the lower connector of the output tube. If 4 mL is not withdrawn from the bioreactors, the difference can be taken from the reservoir.
3. Remove two samples and 4 mL of the medium from the static control.
4. Turn the pump on again and repeat steps 1–4.

3.3.11 *RNA Analysis*

3.3.11.1 RNA extraction

1. Remove the slides from the chambers or wells and wash them three times with cold PBS.
2. Collect the cells in 500 μL Trizol reagent (Invitrogen) and store them at −80°C before RNA extraction.
3. Add 100 μL chloroform (Fisher Scientific, Illkirch, France), cap the tube, and vortex 15 seconds two times.
4. Wait for 2–3 min and centrifuge at 12,000 g for 15 min at 4°C.
5. Add 1 μg of glycogen (Invitrogen) in a new, clean tube (during the centrifuge step).
6. At the end of the centrifuge, transfer the aqueous phase (upper colorless aqueous phase) into the new tube containing glycogen (step 5).
7. Add the same volume of isopropanol (Fisher Scientific) and gently mix by inverting.
8. Wait for 10 min at room temperature.
9. Centrifuge at 12,000 g for 10 min at 4°C.
10. Remove the supernatant and wash the RNA pellet once with 1 mL of 75% cold ethanol (Fisher Scientific).
11. Centrifuge at 12,000 g for 5 min at 4°C.
12. Remove the supernatant and air-dry the RNA pellet (10–20 min at room temperature, cap open).

13. Dissolve the RNA extracted in 15 µL of water, molecular biology grade (Invitrogen).
14. Measure absorbance at 260 nm and determine RNA concentration.
15. Store the RNA at −80°C before reverse transcription (RT).

3.3.11.2 mRNA reverse transcription

1. First, 1 µg of total RNA is retrotranscripted using the RT-MMLV enzyme and random hexanucleotide priming amorces (Invitrogen). In a clean tube, an amount of RNA is added to have 1 µg and completed to 10 µL with water, molecular biology grade (Invitrogen).
2. To perform the RT reaction, 4 µL of RT-MMLV buffer, 2 µL of 0.1 M DTT, 1 µl of dNTPs 10 mM, 1 µL of 0.1 µg/µL random hexanucleotides, 1 µL of RNaseOut, and 1 µL of RT-MMLV (all products from Invitrogen) are added.
3. The thermal cycles are performed in a Mastercycler gradient apparatus (Eppendorf, Pecq, France). Conditions are 37°C for 50 min, followed by 15 min at 70°C. Before the quantitative reverse transcription–polymerase chain reaction (qRT-PCR), dilute complementary deoxyribonucleic acid (cDNA) at 1/10 in water (molecular biology grade, Invitrogen).

3.3.11.3 Quantitative polymerase chain reaction

1. Gene expression is quantified by a real-time PCR assay using Roche LightCycler480 apparatus (Roche Diagnostics, Meylan, France) and LightCycler480 SYBR Green I Master mix (Roche Diagnostics).
2. PCR amplifications are performed with 2 µL of purified cDNA in a 10 µL reaction mixture containing 5 µL of LightCycler480 SYBR Green I Master mix (Roche Diagnostics), 0.05 µL of mix primers 50 µM (Invitrogen) and 2.95 µL of water, molecular biology grade (Invitrogen).

A process has been created on Epmotion5070 robot (Eppendorf, Pecq, France) to distribute samples and reaction mix volumes. Sequences of primers used are listed in Table 3.1.

Table 3.1 Sequence of primers for qRT-PCR

Gene	Sense	Reverse
AAT	TGCTGGGGCCATGTTTTTAG	GGGACCAGCTCAACCCTTCT
AhR	TGGACAAGGAATTGAAGAAGC	AAAGGAGAGTTTTCTGGAGGAA
Albumin	TGCCTGCCTGTTGCCAAAGC	TTGGCAAGGTCCGCCCTGTC
ApoF	GGAAGCGATCAAACCTACCA	ATCAGCCTGACAACCAGCTT
ApoH	GCACTGAGGAAGGAAAATGG	GGCCATCCAGAGAATATCCA
ARNT	TGGCCAAGTCTCGGGTTCCA	TTTGCTGCGCTTGCATCCAT
CAR	CCGTGTGGGGTTCCAGGTAG	CAGCCAGCAGGCCTAGCAAC
C/EBPa	AGGGACTTGGGGCTTGGAAC	GACCCCCATCGCAGTGAGTT
CPS1	TGTCCATTGGTCAGGCTGGA	GCCACCCATGCCCAGAATTA
CYP1A1	TCCGGGACATCACAGACAGC	ACCCTGGGGTTCATCACCAA
CYP1A2	CATCCCCCACAGCACAACAA	TCCCACTTGGCCAGGACTTC
CYP2B6	ATGGGGCACTGAAAAAGACTGA	AGAGGCGGGGACACTGAATGAC
CYP2C9	TCCTATCATTGATTACTTCCCG	AACTGCAGTGTTTTCCAAGC
CYP2D6	CCTACGCTTCCAAAAGGCTTT	AGAGAACAGGTCAGCCACCCACT
CYP3A4	GCCTGGTGCTCCTCTATCTA	GGCTGTTGACCATCATAAAAG
Coagulation factor V	GTGCCCATGACCACATCAGC	CCTCTTCATGTGCCGCCTCT
Coagulation factor VII	TTCAAGGACGCGGAGAGGAC	TGGGGTTTGCTGGCATTTCT
G6P	CGTGATCGCAGACCTCAGGA	GGCTCCCTGGTCCAGTCTCA
GK	TTCGTGTCGCAGGTGGAGAG	GAAGCTGGGGTGCAGCTTGT
GR	CCAACGGTGGCAATGTGAAA	CCGCCAGAGGAGAAAGCAAA
GSTA	TGGCAGAGAAGCCCAAGCTC	TGCACCAGCTTCATCCCATC
HNF4a	CGCAGATGTGTGTGAGTCCA	CAGTGCCGAGGGACAATGTA
MDR1	GACCGGACATCCCAGTGCTT	TGTGCTCGGAGCCACTGAAC
MRP2	GGCCCCAATGGCCTGTTGAA	TGGAGGCTGCATCTTCGGGG
NTCP	TCACCCTGCCACCCAACTTT	GACCTTGCCCAGCACAAAGG
OATP8	GGTCCAGTCATTGGCTTTGC	ACAAGGAAACCAAGCCACCA
PEPCK 1	CGGCATCGAGCTGACGGATT	CGCCGTACCCACTGCCAAAG
PK-L	ACCGGCAGCTGTTTGAGGAG	CAGAGCGGGTGACAGCAATG
PXR	GGACCAGCTGCAGGAGCAAT	CATGAGGGGCGTAGCAAAGG
RPLP0	TCGACAATGGCAGCATCTAC	GCCTTGACGTTTTCAGCAAG
UGT1A1	GGTGACTGTCCAGGACCTAT	TAGTGGATTTTGGTGAAGGCAGTT
UGT2B4	CTTTAGGACTCAATACTCGGCTG	CTCATAGATGCCATTGGCTCCAC
UGT2B7	CCTTAGGTCTCAATACTCGGCTG	CTCGTAGATGCCATTGGCTCCA

Abbreviations: AAT, α1-antitrypsin; AhR, aryl hydrocarbon receptor; ARNT, aryl hydrocarbon receptor nuclear translocator; Apo, apolipoprotein; CAR, constitutive androstan receptor, C/EBP CAAT, box enhancer binding protein; CPS-1, carbamoyl phosphate synthase-1; CYP, cytochrome P450; G6P. glucose-6-phosphatase; GK, glucokinase; GR, glucocorticoid receptor; GST, glutathion-S-transferase; HNF, hepatocyte nuclear factor; MDR, multidrug resistance; MRP, multidrug resistance protein; NTCP, Na/taurocholate cotransporting polypeptide; OATP, organic anion transporter; PEPCK, phosphoenolpyruvate carboxykinase ; PK-L, pyruvate kinase isoform L; PXR, pregnane X receptor, RPLP0; ribosomal protein, large, P0; UGT, UDP-glucuronosyltransferase.

Table 3.2 Reference substrates and specific enzymatic reactions catalyzed by human CYP2C9, 2D6, and 3A4

CYP	Marker substrate	Enzymatic reaction	Specific metabolite quantified
CYP2C9	Tolbutamide	Methyl-hydroxylation	4-methylhydroxy-tolbutamide
CYP2D6	Dextromethorphan	*O*-Demethylation	Dextrorphan
CYP3A4	MDZ	1′-Hydroxylation followed by conjugation	1′-OH-MDZ followed by 1′-OH-MDZ-Glu

Abbreviations: 1′-OH-MDZ, 1′-Hydroxy-midazolam; 1′-OH-MDZ-Glu, 1′-Hydroxy-midazolam glucuronidation.

The following program is suggested: one step at 95°C for 10 min, 55 cycles of denaturation at 95°C for 15 seconds, annealing at 65°C for 15 seconds, and elongation at 72°C for 15 seconds. Amplification specificity/quality is assessed by analyzing product-melting curves.

Relative quantification was calculated with the LightCycler480 software (Roche Diagnostics) and normalized to ribosomal protein, large, P0 (RPLP0) RNA expression. Whenever possible, sequences of primers are designed from different exons to avoid false positives due to DNA contamination.

3.3.12 *Characterization of Drug Metabolism Capacity*

1. For the drug metabolism experiment coverslips with hepatocytes are placed in 24-well plates. The medium is renewed with 180 μL of fresh culture medium. Add CYP isoforms probes (see Table 3.2) directly.
2. A 20 mM stock solution of dextromethorphan (Sigma) and 5 mM stock solutions of midazolam (MDZ) (Sigma) and tolbutamide (Sigma) in Dimethylsulfoxide (DMSO) (i.e., 1000X stock solutions) are prepared.
3. These solutions are diluted 100-fold in the culture medium containing 1% (w/v) bovine serum albumin (BSA) in order to obtain 10X working solutions.
4. Then 20 μL of each working solution of each CYP isoform probe is added to the 180 μL medium already present in the wells in order to achieve a 5 μM final concentration for MDZ and tolbutamide—20 μM for dextromethorphan and a 0.1% (v or w/v) final concentration of solvent and BSA.

Table 3.3 Suggested LC-MS/MS analytical conditions for substrates and metabolites for drug metabolism assay

Compound	Formula	Mass transitions	Cone voltage (V)	Collision energy (eV)	Ionization mode
Dextromethorphan	Dextromethrophan $C_{18}H_{25}NO$	272.2 > 147.2	30	32	ES^+
	Dextrorphan $C_{17}H_{23}NO$	258.2 > 157.2			
MDZ	MDZ $C_{18}H_{13}ClFN_3$	326.1 > 291.2		28	
	1′-OH-MDZ $C_{18}H_{13}ClFN_3O$	342.2 > 324.2	30	22	ES^+
	1′-OH-MDZ-Glu $C_{24}H_{22}ClFN_3O_7$	518.1 > 324.1		25	
Tolbutamide	Tobutamide $C_{12}H_{18}N_2O_3S$	269.2 > 170.1	30	16	ES^-
	4-OH-tolbutamide $C_{12}H_{18}N_2O_4S$	285.1 > 186.1			

5. For the determination of the metabolism of different probes, kinetic studies over 24 h for dextromethorphan, MDZ, and tolbutamide (sampling times: 2, 4, 8, and 24 h).
6. At each selected time point, 400 μL of an acetonitrile/water mixture (30/10, v/v) (Sigma) is added to each specific well, and both the extracellular medium and the cell compartment are scraped and mixed together. The cell homogenate can be transferred to a glass test tube and stored frozen at −20°C until further analysis.
7. Prior to analysis, cell homogenates are sonicated for a few seconds, homogenized, and centrifuged at 3,000 rpm for 20 min. After sample transfer to a 96-well plate and threefold dilution, supernatants are analyzed for unchanged drugs and specific metabolites by liquid chromatography/mass spectrometry (LC/MS) (Table 3.3): MDZ, 1′-hydroxy-midazolam (1′-OH-MDZ) (Sigma), and 1′-hydroxy-midazolam-glucuronide (1′-OH-MDZ-Glu) (synthesized by the Isotope Chemistry and Metabolites Department of Sanofi Recherche, Chilly-Mazarin, France), dextromethorphan and dextrorphan (Sigma), tolbutamide, and 4-hydroxy-tolbutamide (Sigma).
8. The data are collected and processed using MassLynx 4.0 (Waters-Micromass, Milford, MA, USA). The chromatograph was fitted with a C_{18} Hypersil BDS® column (125 × 3.0 mm i.d., 3 μm particle size [Agilent]).

9. The mobile phase was a mixture of 1.5 g/L ammonium acetate–2 mL/L formic acid (solvent A) and acetonitrile 80%–methanol 20%–0.15g/L ammonium acetate–formic acid 2 mL/L (solvent B). The solvent programmer was set to deliver a flow rate of 0.25 mL/min. Compounds were eluted with a linear gradient from 10% to 90% solvent B over 1.5 min, followed by an isocratic step at 90% for 3 min.
10. Results are expressed as the initial velocity of metabolite formation respective to each isoform in nmol/h/10^6 hepatocytes.

References

1. Krewski, D., Westphal, M., Al-Zoughool, M., Croteau, M. C., Andersen, M. E. (2011) New directions in toxicity testing. *Annu Rev Public Health* **32**, 161–178.
2. Vinci, B., et al. (2011) Modular bioreactor for primary human hepatocyte culture: medium flow stimulates expression and activity of detoxification genes. *Biotechnol J* **6**, 554–564.
3. Gómez-Lechón, M. J., Lahoz, A., Gombau, L., Castell, J. V., Donato, M. T. (2010) In vitro evaluation of potential hepatotoxicity induced by drugs. *Current Pharm Design* **16**, 1963–1977.
4. Pichard, L., Raulet, E., Fabre, G., Ferrini, J. B., Ourlin, J. C., Maurel, P. (2006) Human hepatocyte culture. *Methods Mol Biol* **320**, 283–293.
5. Berthiaume, F., Moghe, P. V., Toner, M., Yarmush, M. L. (1996) Effect of extracellular matrix topology on cell structure, function, and physiological responsiveness: hepatocytes cultured in a sandwich configuration. *FASEB J* **10**, 1471–1484.

Chapter 4

Generation of Patient-Specific Cardiac Patches by Human Cardiac Progenitor Cells and 3D Scaffolds

Giancarlo Forte

Biomaterials Unit, International Center for Materials Nanoarchitectonics (MANA), National Institute for Materials Science (NIMS), 1-1 Namiki, Tsukuba 305-0044, Japan

forte.giancarlo@nims.go.jp

Human cardiac progenitor cells (hCPCs) can be harvested from human atrial specimens during routine bioptic analysis. These cells can be purified using the MACS magnetic system by means of the Sca-1 antibody and easily cultured in standard culture conditions. Although the Sca-1 antigen (also known as Ly6A/E or Ly6D) is known to be exclusively expressed in murine cells, the use of antibodies directed against it allows for the purification of a homogeneous population of human progenitor cells. hCPCs retain their plasticity in culture and proliferate without the occurrence of spontaneous events of differentiation. When implanted in the murine heart in vivo, they can integrate in the muscle and vascular tissue texture, thus suggesting their ability to participate in cardiac repair. In the present chapter we describe the technique to isolate and culture hCPCs and the possibility to use three-dimensional scaffolds having controlled porosity and anisotropic mechanical

Cellular in vitro Testing: Methods and Protocols
Edited by John W. Haycock, Arti Ahluwalia, and John M. Wilkinson

ISBN 978-981-4364-97-3 (Hardcover), 978-981-4364-98-0 (eBook)
www.panstanford.com

properties. The scaffolds would provide physical support to the cells, while favoring their alignment and eventually a certain degree of precommitment. Moreover efficient oxygen supply to the core of the constructs as well as catabolite removal will be provided. This strategy is proposed to generate patient-specific cardiac substitutes for future clinical application.

4.1 Introduction

Cardiac progenitor cells (CPCs) are defined as self-renewing cardiac tissue–resident stem cells retaining clonogenic ability and the capacity to generate the main components of cardiac tissue: cardiomyocytes, endothelial cells, and smooth muscle cells. Their existence has been denied for a long time in view of the postulated postmitotic status of the heart and has been compellingly demonstrated in the last years, showing that they are responsible for cardiac tissue homeostasis: at the age of 25 years, no more than 1% of the cardiomyocytes are annually substituted by progenitor cells, this percentage reducing to less than 0.5% at the age of 75 [1]. The presence of a small reservoir of cardiac resident progenitor cells (CPCs or cardiac stem cells [CSCs]) within cardiac muscle has been for the first time described in an in vivo investigation by the group of Piero Anversa in 2002, who demonstrated that undifferentiated cells from the recipient could contribute to donor cardiac muscle and vessel formation in sex-mismatched heart transplants [2]. Although the nature of such cells could not be assessed by that approach, the results paved the way to a number of investigations that indeed identified undifferentiated cells in the hearts of rodents [3], dogs [4], and humans [5]. Their stem cell status was at first associated to the expression of the c-kit (CD117) antigen, a marker already in use in clinical practice to recognize undifferentiated cells within tumors. Along with CD117-expressing cells, other classes of cardiac progenitors can be recognized by the expression of multidrug resistance-1 (MDR-1), stem cell antigen-1 (Sca-1), or Isl-1 (Islet-1) [6].

The debate on the nature of these cardiac cells is still open, while all the subsets so far identified through the expression of stemness markers (c-kit+, Sca-1+, Islet-1+) demonstrated the ability to give birth to new contractile cells in vitro. Interestingly, only c-kit+ and Sca-1+ progenitors were shown to be involved in postnatal cardiac tissue homeostasis in vivo [7], the presence of Islet-1+ cells being limited to fetal life. However, the very low number of c-kit+ cells in the myocardium is considered a major limiting factor for their use in cardiac regeneration [8], the expression of such marker being probably restricted to a limited number of CSCs. The expression of a Sca-1-like marker in rodent [9] and even human cells appears to be more stable and reliable instead [10]. Sca-1+ cells can be easily extracted and purified from the adult mouse and human heart, and their proliferation in culture is robust and their plasticity preserved.

Multipotent Sca-1+ cells are represented in a limited number in the myocardium [9], where they contribute to cardiac tissue homeostasis [10]. These cells express low amounts of hematopoietic lineage markers (c-kit, CD34, CD45). They can also express markers for mesenchymal progenitors (CD105, nestin, nucleostemin) as well as early cardiac markers (GATA-4, Nkx-2.5, and MEF-2c) and proceed to cardiomyogenic differentiation in vitro when treated with FGF-2 [9], TGF-β1 [12], and 5-azacytidine [13]. Sca-1 function has been correlated to CPC proliferation and maintenance as well as in response to hypoxic and ischemic conditions. In fact, Sca-1+ CPCs colonize the infarct border zone and generate new cardiomyocytes when systemically injected in infarcted mice [14]. Consistently, in Sca-1 knockout mice the functional improvement of the damaged myocardium after CPC transplantation is attenuated and the formation of new vessels and the inhibition of myocardial apoptosis severely impaired [11].

The successful application of tissue engineering techniques to cardiac repair relies on the possibility to obtain a meaningful number of cells able to produce new cardiomyocytes in a controlled and timely fashion. These cells can be cultured onto three-dimensional (3D) scaffolds to prepare solid cardiac constructs in vitro before implantation (Fig. 4.1). As a principle, the cardiac construct produced should be able to comply with cardiac muscle architecture and be deformable enough to indulge and sustain cardiac contraction. Moreover, it should be able to preserve cardiac tissue integrity, while favoring stem cell electromechanical coupling

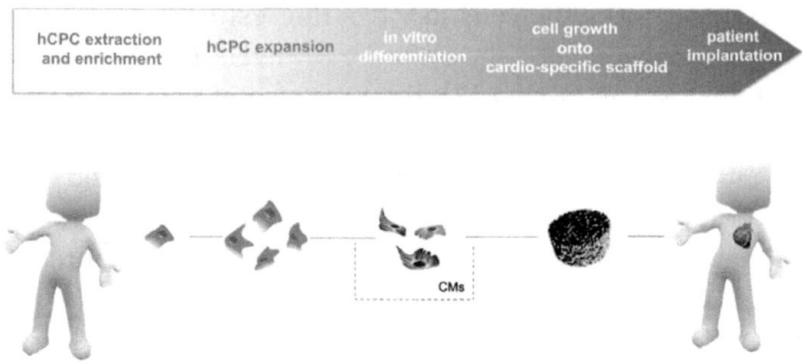

Figure 4.1 hCPCs can be extracted from patient biopsies, amplified in culture, and seeded onto 3D porous scaffolds to produce thick cardiac substitutes to be implanted in vivo. Before seeding the cells on the scaffold, a preliminary differentiation step by means of biochemical cues can be performed.

with the host tissue. Finally, cardiac-specific scaffolds are also required to promote newly formed tissue vascularization.

4.2 Materials

4.2.1 *Cardiac Biopsies Harvesting and Handling*

1. Human Sca-1+ CPCs can be isolated from human auricular biopsies [11, 12]. The protocol described hereafter is a modification of the one utilized to isolate Sca-1+ CPCs from murine heart and described elsewhere [15].
2. Auricular specimens (1–3 cm^3) are harvested and kept in ice-cold sterile phosphate buffered saline (PBS) and treated under a sterile laminar flow hood. The specimens are washed several times with sterile, ice-cold PBS and then mechanically minced by a sterile scalpel.
3. Samples are incubated at 37°C for 2 h in 0.05% trypsin, 0.02% ethylenediamine tetra-acetic acid (EDTA) in PBS, followed by collagenase II (1500 U, Worthington Biochemical Corporation) digestion in Leibovitz medium (Worthington) for 60 min at 37°C. The digestion is blocked by adding 2 volumes of the cold medium.

4. Tissue fragments are centrifuged at 4°C for 10 min at 800xg and the pellets resuspended in Dulbecco's modified Eagle's medium (Lonza Group Ltd.), 10% fetal bovine serum (FBS; Lonza), 100 IU/mL penicillin and 100 μg/mL streptomycin, insulin-transferrin-selenium 1X, retinoic acid 300 ng/mL, linoleic acid 0.8 μg/mL, L-glutamine 2 mM, insulin-like growth factor 1 0.1 ng/mL, and endothelial growth factor 0.1 ng/mL. Therefore the fragments are exposed to filtration through a 100 μm cell strainer (BD Biosciences) and incubated at 37°C, 5% CO_2 in a complete medium.
5. The following day, medium- and nonadhering fragments are removed and replaced with a fresh complete medium, which is changed every second day. After 10–15 days, fibroblastoid cells will migrate from the fragments and reach confluence. Such cells can be expanded in a complete medium as a heterogeneous population of nonmiocytes or purified to obtain Sca-1+ cells.

4.3 Methods

4.3.1 *Isolation and Purification of Human Cardiac Progenitor Cells by Sca-1-Like Antigen Expression*

1. The cells will migrate out of the tissue fragments and give raise to colonies. At this stage, cells can be detached by trypsin solution (0.05% trypsin, 0.02% EDTA in PBS) and cloning rings to obtain monoclonal cell populations, which contain Sca-1+ cells.
2. Otherwise, adherent cells can be grown up to confluence, detached, and passaged at this stage. In both cases, the cells can be harvested by trypsin solution and exposed to magnetic sorting by MACS technology. The cells are washed in PBS and centrifuged twice for 8 min at 300×g.
3. The pellet is resuspended in a buffer solution (PBS pH 7.2, 0.5% bovine serum albumin [BSA], 2 mM EDTA). Up to 10^7 cells can be resuspended in 100 μL buffer.
4. The Anti-Sca-1-FITC antibody (Miltenyi Biotec GmbH) is added at a concentration of 1:10. The cells are incubated for 10 min

in the dark at 2–8°C, washed with 1–2 mL of buffer, and centrifuged for 10 min at 300×g.

5. The pellet should be resuspended in 80 μL of buffer per 10^7 total cells and 20 μL of anti-FITC microbeads added to the suspension. After mixing vigorously, incubate for 15 min at 2–8°C. After preparing the elution column by buffer equilibration, it is possible to proceed to magnetic separation.
6. The Sca-1+ cell fraction can be plated on fibronectin (2 μg/mL, Sigma-Aldrich) precoated dishes at a concentration of 2.0×10^4 cells/cm^2 in a complete medium and used for the following experiments.

4.3.2 *Characterization of Human Cardiac Progenitor Cells*

1. When cultured in a proliferation medium, human cardiac progenitor cells (hCPCs) display high positivity for the Sca-1 marker (95–98%) and limited positivity for c-kit antigen (10–15%), CD90 (15–30%), and CD105 (30–50%).
2. The cells do not express the hematopoietic marker CD34 and are hardly stained with cardiomyoblast markers like GATA-4 (3–7%) and Nkx-2.5 (5–20%). Endothelial cells are also scarcely represented within this population (FLK1: 4–8%, CD31: not detectable), while spontaneous cardiac differentiation is extremely rare (Fig. 4.2).
3. Cardiac differentiation of human Sca-1+ CPCs can be obtained by direct coculture with neonatal cardiomyocytes [11] or by stimulation with TGF-β in the presence of 5-azacytidine [12]. By both protocols, the de novo expression of sarcomeric structural

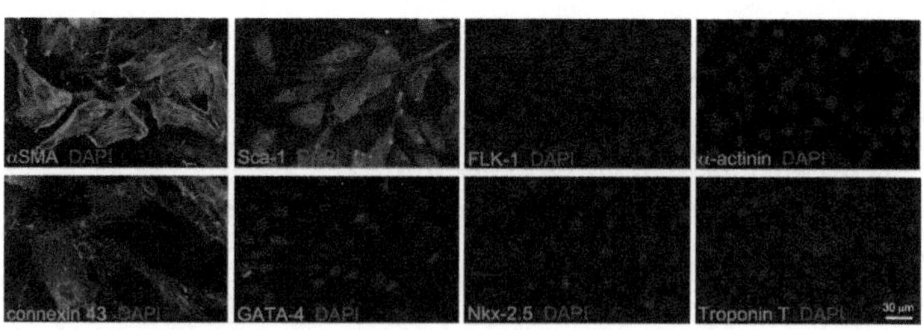

Figure 4.2 The human Sca-1+ CPC phenotype in culture is relatively stable.

proteins and of the functional adherens junction and desmosomal proteins has been demonstrated. It is interesting to note that the mature electrical properties of differentiated hCPCs have been shown [16].

4.3.3 *Sca-1+ Cardiac Progenitor Cell Differentiation by Neonatal Cardiomyocyte Coculture*

1. Neonatal cardiomyocytes can be extracted from the hearts of 1–3-day-old C57/Bl/6 mice, following the manufacturer's instructions (Worthington Biochemical Corp.), and plated onto fibronectin (2 μg/mL), laminin (0.2%), and gelatin (0.02%, Sigma-Aldrich) precoated glass chamber slides.
2. After 24 h, Sca-1+ CPCs prestained with the viable red fluorescent dye Vybrant DiI (Molecular Probes), following the manufacturer's specifications, are seeded directly onto cardiomyocytes (ratio 1:5) in a complete medium.
3. The cells are cocultured for seven days before being processed for immunofluorescence analysis using cardiac-specific antibodies (Fig. 4.3).

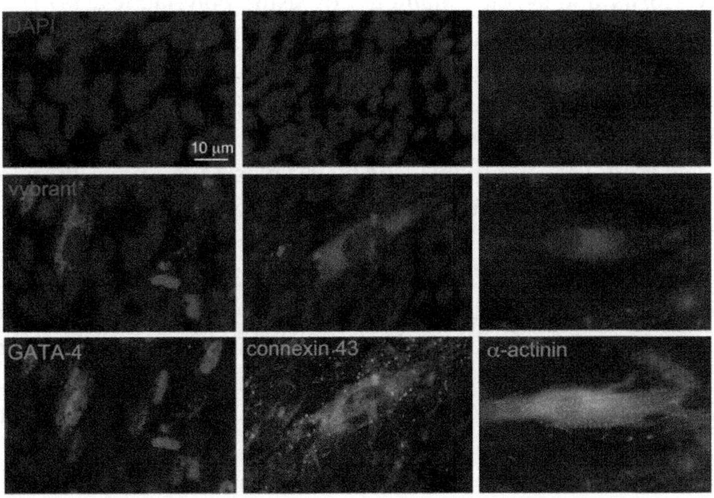

Figure 4.3 Human Sca-1+ CPCs can be differentiated by murine neonatal cardiomyocyte coculture for seven days.

4.3.4 *Culture of Human Cardiac Progenitor Cells onto 3D Scaffolds with Controlled Porosity*

1. Cardiac-specific scaffolds have to comply with a number of requirements; in fact, myocardial contractile function relies on the transmission of electrical and mechanical forces throughout a functional syncytium. So, the integrity of the tissue has to be preserved. Moreover, as far as stem cell engraftment is concerned, scaffolds should be able to start at least cell alignment and a commitment to favor stem cell electromechanical coupling with the host tissue. The pressure-assisted microsyringe (PAM) is a microfabrication technique developed and patented by the University of Pisa, allowing accurate design of 3D scaffolds having controlled porosity and anisotropic mechanical properties distribution. Using such a technique, the polymer solution is extruded from a microsyringe mounted on a three-axis micropositioner applying a pressure generated by compressed air. The system is interfaced to a personal computer with a built-in graphics tool for creating or importing scaffold designs.
2. Scaffold geometry is controlled by modulating the applied pressure and deposition speed and is also a function of polymer viscosity and syringe diameter. By using poly(lactic acid) (PLA), it is indeed possible to produce square grids with unit sides of 200 μm, and line widths of 50 μm, fabricated by depositing three layers. Each layer is laterally offset with respect to adjacent layers by 100 μm [17].
3. The scaffolds can be sterilized by gas plasma or under ultraviolet (UV) light for 15 min. Finally, three washes in 70% ethanol under a sterile bench are recommended. After washing, the scaffolds can be rinsed twice with sterile PBS and then transferred to an ultralow attachment culture plate (Corning Inc.) in a complete growth medium at 37°C for 6–8 h. Should the scaffolds float during cell culture, a Teflon ring or support (e.g., Scaffdex Cell Crown, Tampere, Finland) may be used to let them soak in the medium.
4. Human Sca-1+ CPCs can be harvested as previously described, suspended in a complete growth medium and seeded at a concentration of 2.0–3.0 $\times$ 10^4 cells/cm^2. The growth medium will be changed every second day (Fig. 4.4).

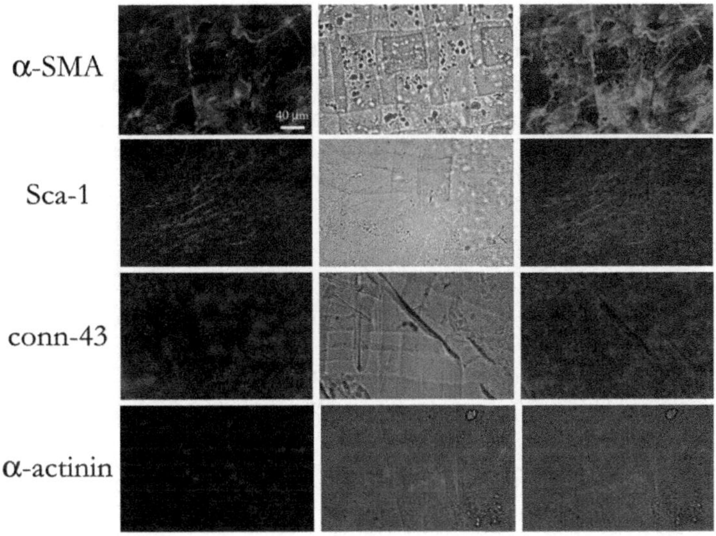

Figure 4.4 Human Sca-1+ CPCs can be cultured onto 3D PLA scaffolds with square pores up to seven days in the absence of spontaneous events of differentiation.

4.4 Notes

1. Written consent of the donor and previous approval by the ethic committee of the institution are required to handle human cells.
2. The procedures for human biopsy handling and CPC isolation are performed in ice, unless otherwise stated.
3. Slight changes in the time points of cell migration out-of-tissue fragments can occur.
4. Continuous stirring of the cell suspension during Sca-1 antibody incubation can enhance the yield of cell purification.
5. Sca-1+ CPCs retain their phenotype and plasticity over a number of passages in culture, but repeated purification with Sca-1 microbeads may be necessary from time to time.
6. Ex vivo–extracted cells can display donor-to-donor variability in terms of phenotype, proliferation, and differentiation.
7. The use of cellular and tissue-based products in human disease therapy is subjected to regulations by the European Union (EU) and the Food and Drug Administration (FDA), aimed at

establishing classification criteria for advanced therapy medicinal products (ATMPs). In particular, the European Regulation states that human cells to be used in cell therapy have to comply with the Good Manufacturing Practice (GMP) protocols.

References

1. Bergmann, O., et al. (2009) Evidence for cardiomyocyte renewal in humans. *Science* **324**, 98–102.
2. Quaini, F., et al. (2002). *N Engl J Med* **346**, 5–15.
3. Urbanek, K., et al. (2006) Stem cell niches in the adult mouse heart. *Proc Natl Acad Sci USA* **103**, 9226–9231.
4. Linke, A., Müller, P., Nurzynska, D., Casarsa, C., Torella, D., Nascimbene, A. (2005) Stem cells in the dog heart are self-renewing, clonogenic, and multipotent and regenerate infarcted myocardium, improving cardiac function. *Proc Natl Acad Sci USA* **102**, 8966–8971.
5. Quaini, F., et al. (2004) The regenerative potential of the human heart. *Int J Cardiol* **95**(Suppl 1), S26–S28.
6. Laugwitz, K. L., et al. (2005) Postnatal isl1+ cardioblasts enter fully differentiated cardiomyocyte lineages. *Nature* **10**(433), 647–653.
7. Bearzi, C., et al. (2007) Human cardiac stem cells. *Proc Natl Acad Sci USA* **10**, 14068–14073.
8. Di Nardo, P., Forte, G., Ahluwalia, A., Minieri, M. (2010) Cardiac progenitor cells: potency and control. *J Cell Physiol* **224**, 590–600.
9. Rosenblatt-Velin, N., Lepore, M. G., Cartoni, C., Beermann, F., Pedrazzini, T. (2005) FGF-2 controls the differentiation of resident cardiac precursors into functional cardiomyocytes. *J Clin Invest* **115**, 1724–1733.
10. Tateishi, K., et al. (2007) Clonally amplified cardiac stem cells are regulated by Sca-1 signalling for efficient cardiovascular regeneration. *J Cell Sci* **120**, 1791–1800.
11. Forte, G., et al. (2011) Human cardiac progenitor cell grafts as unrestricted source of super-numerary cardiac cells in healthy murine hearts. *Stem Cells*, **29**(12), 2051–2061.
12. Smiths, A. M., van Vliet, P., Metz, C. H., Korfage, T., Slujiter J. P., Goumans, M. J. (2009) Human cardiomyocyte progenitor cells differentiate into functional cardiomyocytes: an in vitro model for studying cardiac physiology and pathophysiology. *Nat Prot* **4**, 232–243.

13. Oh, H., et al. (2003) Cardiac progenitor cells from adult myocardium: homing, differentiation, and fusion after infarction. *Proc Natl Acad Sci USA* **100**, 12313–12318.
14. Wang, X., et al. (2006) The role of Sca-1+/CD31- cardiac progenitor cell population in post-infarction left ventricular remodeling. *Stem Cells* **24**, 1779–1788.
15. Forte, G., et al. (2008) Criticatility of the biological and physical stimuli array inducing resident stem cell determination. *Stem Cells* **26**, 2093–2103.
16. de Boer, T. P., et al. (2010) Human cardiomyocyte progenitor cell-derived cardiomyocytes display a maturated electrical phenotype. *J Mol Cell Cardiol* **48**, 254–260.
17. Mariani, M., Rosatini, F., Vozzi, G., Previti, A., Ahluwalia, A. (2006) Characterization of tissue-engineered scaffolds microfabricated with PAM. *Tissue Eng* **12**, 547–557.

Chapter 5

In vitro Risk Assessment of Nanoparticles

Birgit K. Gaiser,[a] Julia Susewind,[b] Nadia Ucciferri,[c] Eva-Maria Collnot,[d] Arti Ahluwalia,[c] and Vicki Stone[a]

[a] *Heriot-Watt University, School of Life Sciences, Nanosafety Research Group, Edinburgh, U.K.*
[b] *Department of Biopharmaceutics and Pharmaceutical Technology, Saarland University, Saarbrücken, Germany*
[c] *Interdepartmental Research Center "E.Piaggio", Faculty of Engineering, University of Pisa, Via Diotisalvi, 2 - 56126 Pisa, Italy*
[d] *Department of Drug Delivery, Helmholtz Institute for Pharmaceutical Research Saarland, Saarland University, Saarbrücken, Germany*
b.gaiser@hw.ac.uk

With the increasing production and use of nanoparticles, the likelihood of human exposure, either accidental or intentional, and as consumers or in the occupational setting, is also increasing. It is therefore paramount to ensure the safety of these novel materials and allow for restrictions to be placed on their use, where applicable, or take relevant measures to reduce exposure. Nanoparticles present special challenges to in vitro risk assessment, which relate to their small size and resulting properties, such as increased surface area and reactivity, their behavior in solution, the need for extensive physicochemical characterization, and their

Cellular in vitro Testing: Methods and Protocols
Edited by John W. Haycock, Arti Ahluwalia, and John M. Wilkinson

ISBN 978-981-4364-97-3 (Hardcover), 978-981-4364-98-0 (eBook)
www.panstanford.com

interaction with a number of assays. These features necessitate the implementation of additional controls in many assays and render some standard assays impracticable. This chapter comprises a set of standard operating protocols used by the partners of the FP7 project InLiveTox to give a brief overview of testing for possible adverse effects of nanoparticles in vitro, using methods applicable for most adherent cells. The protocols include culture of human endothelial cells, intestinal epithelial cells, and hepatocytes; preparation of nanoparticle suspensions; assessment of particle size distribution; cytotoxicity and viability assays; testing of specific cellular functionality in intestinal epithelial cells, endothelial cells, and hepatocytes; quantification of markers of inflammation on the RNA and protein level; and measurement of reduced glutathione (GSH) levels as a measure of oxidative stress.

5.1 Introduction

5.1.1 *Nanoparticles and Nanomaterials*

Materials with at least one dimension measuring 100 nm or less are referred to as nanomaterials and include nanofilms (one dimension $\leq$100 nm), nanofibres and nanorods (two dimensions $\leq$100 nm), and nanoparticles (NPs; all dimensions $\leq$100 nm [1]. Particles on the nanoscale can occur naturally, for example, soot particles after combustion processes or large organic molecules [2]. In the last few decades, however, manufacture and application of NPs have become a new and increasingly important branch of industry [3]. The small size of NPs has various implications in their properties when compared to larger-size particles: There is an increased surface area per mass unit of material, and therefore a larger percentage of the atoms are present on the surface (Fig. 5.1). In addition to the larger surface area, the decreased stability of bonds in the smaller-size particles increases the surface reactivity [4].

NPs have become promising tools in many applications and are added to many consumer products: Nanosilver is used for its antibacterial properties, for example, in medical products [5]; nanometal oxides are used in paints and cosmetics such as

Particle	A	B	C
Diameter	d	0.1 d	0.01 d
Particles per mass unit	n	1,000 n	1,000,000 n
Surface area per mass unit	A	10 A	100 A

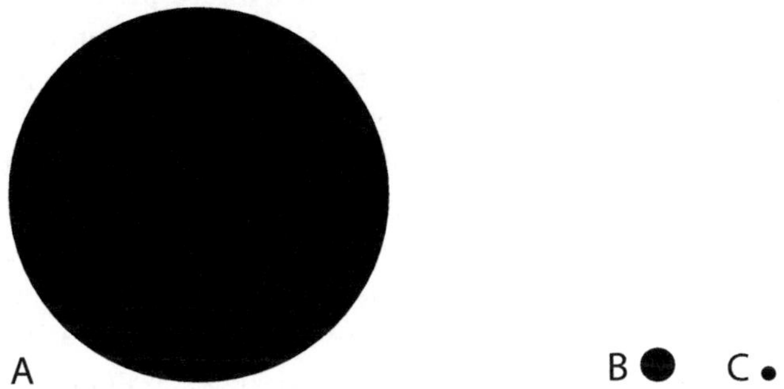

Figure 5.1 Spheres of decreasing size and the relationships between their diameters and surface areas. As the particle size decreases, the number of particles per mass unit increases by the cube of the size difference factor (i.e., a 1,000-fold increase for a 10-fold decrease in size), and the surface area per mass unit, along with the percentage of atoms in the material being present on the particle surface, increases by the size difference (i.e., a 10-fold increase for a 10-fold decrease in size).

sunscreens [6, 7]; and various nanomaterials are candidates for medical applications such as imaging and drug delivery [8, 9].

5.1.2 *Nanotoxicology*

With the increase in the manufacture of NPs and NP-containing products, there is an increased risk of occupational exposure and also an increased risk of intentional exposure, for example, to medical NPs and NPs in cosmetics and personal care products. The changed properties of NPs compared to larger, or "bulk," particles of the same material mean that their effects on the human body and the environment can be more severe than the effects of an equal mass dose of larger particles [4]. It has been suggested that for particles generally associated with low toxicity, the surface area dose is a better predictor of their adverse effects than their mass, meaning toxicity and inflammogenicity of the same mass of particles increase with decreasing particle radius [10]. Another concern is that NPs may cross barriers that larger particles are unable to cross, and indeed transition of NPs from the lungs and gastrointestinal (GI) tract into the bloodstream have been observed [11, 12].

On a cellular level, many NPs are of a size where interaction with the deoxyribonucleic acid (DNA) is possible if transition into the nucleus takes place [13], and reactive surfaces as well as surface contamination by metals or organic molecules can cause oxidative damage to membranes and proteins and also potentially the DNA [14]. For these reasons, it is important that NPs be tested comprehensively for potential adverse effects. Regulators have acknowledged this problem in publications and calls for researchers to investigate NP toxicity [15]. InLiveTox (www.inlivetox.eu) is a 7th Framework Programme project, in which a Quasi-Vivo® system [16] will be used to investigate transport of particles through an in vitro model of the GI barrier [17], and effects of transported particles on endothelial cells and hepatocytes connected to the GI model by medium flow will be examined and compared to an in vivo study to evaluate the in vitro model.

Initial experiments focused on establishing protocols for measuring endpoints associated with NP toxicity, such as cytotoxicity, pro-inflammatory gene expression and cytokine release, cellular functionality, and antioxidant depletion. These protocols are summarized in this chapter to provide researchers with a set of procedures for the evaluation of NP toxicity, which can be used for most adherent human cells.

5.1.3 *Importance of Appropriate Controls and Physicochemical Characterization*

Some challenges in in vitro nanotoxicology are different from the ones faced when working with larger particles. For instance, it can be difficult to remove NPs from a cell culture medium by conventional methods such as centrifugation or filtering. Therefore, additional controls, for example, a cell-free medium containing NPs, need to be run to eliminate the possibility of NP interference with assays such as fluorescent or absorbance measurements [18].

Another feature of NPs is that their large surface area can adsorb biomolecules, which can impact protein or cytokine quantification [19]. For these assays, additional controls must include incubating known concentrations of the protein or cytokine, such as a standard curve with NPs, for the duration of the experimental exposure to see if the particles influence the recovery rate.

Lastly, it is essential to characterize the particles to be able to relate their effects to physicochemical characteristics. These can change depending on the cell culture medium, serum concentration, pH, and ionic strength [20]. Standard characterization techniques include size measurement by dynamic light scattering (DLS) or transmission electron microscopy (TEM), determination of the surface area (Brunauer–Emmert–Teller, or BET method), charge measurement (zeta-potential), aggregation state (by DLS), and solubility for metals and some metal oxides (inductively coupled plasma mass spectrometry or optical emission spectroscopy, ICP-MS or ICP-OES. We describe DLS measurement as a widely used and relatively simple characterization technique. However, to understand NP toxicity, more than one technique is generally used.

5.2 Materials

5.2.1 *Cell Culture and General Maintenance of Cells*

5.2.1.1 Human umbilical vein endothelial cells

1. Human umbilical vein endothelial cells (HUVECs) can be obtained from commercial sources (e.g., Promocell, EACC, ICLC) or by collagenase digestion of the umbilical vein. Cells can be used for up to eight to nine passages [21, 22]. The surface cell concentration is 20,000 cells/cm^2 at confluence.
2. Eagle's minimum essential medium (EMEM) supplemented with 1% penicillin/streptomycin/amphotericin B 10K/10K/25 μg, 4 mM L-glutamine, 10% fetal calf serum (FCS), 1% nonessential amino acids 100×, 1% modified Eagle's medium (MEM) vitamins solution 100×, 10 μg/mL endothelial cell growth supplement (ECGS), 10 ng/mL human epidermal growth factor (hEGF) 3 ng/mL basic fibroblast growth factor (bFGF), 1 μg/mL hydrocortisone, and 10 μg/mL heparin sodium salt.

3. Phosphate buffered saline (PBS), pH 7.4.
4. 0.25% trypsin (w/v) and 1 mM ethylenediamine tetraacetic acid (EDTA).
5. 1% gelatin from porcine skin in PBS solution, autoclaved, to coat cell culture surfaces, or gelatin-coated tissue culture plastics such as polystyrene T75 flasks for HUVEC passage.

5.2.1.2 Caco-2 human epithelial colorectal adenocarcinoma cells

1. Caco-2 cells' clones C2Bbe1 Caco were obtained from the American Type Culture Collection (ATCC). Cells can be used for 12–20 passages without change of phenotype.
2. Dulbecco's modified Eagle's medium (DMEM) supplemented with 10% FCS and 1% nonessential amino acids 100x.
3. PBS, pH 7.4.
4. 0.25% trypsin (w/v) and 1 mM EDTA.

5.2.1.3 C3A human hepatoma cells

1. C3A cells are available commercially (e.g., ATCC). C3A cells can be used for 10–20 passages.
2. MEM (e.g., M2279, Sigma) supplemented with 10% FCS, 1% nonessential amino acids 100×, 2 mM L-glutamine, 1% penicillin/streptomycin (10 K/10 μg/mL), and 1 mM sodium pyruvate.
3. PBS, pH 7.4.
4. 0.25% trypsin (w/v) and 1 mM EDTA.
5. Collagen type I from rat tail (e.g., Sigma).

5.2.2 *Particle Preparation and Cell Exposures*

1. Sonicating water bath
2. Precision weighing balance, to be kept in a glove box or similar for respiratory protection when using particles in powder form or fibers
3. NPs to be tested, positive and negative controls for the respective assays
4. Sterile water containing 2% FCS
5. Cell culture medium

5.2.3 *Particle Characterization by Dynamic Light Scattering*

1. Sonicating water bath
2. Precision weighing balance, to be kept in a glove box or similar for respiratory protection when using particles in powder form or fibers
3. Zetasizer Nano, Zetasizer Nano Software 6.20 (Malvern)
4. Nanosight, NTA software 2.0 (NanoSight)
5. Respective NPs to be tested
6. Sterile water containing 2% FCS
7. Cell culture medium

5.2.4 *Toxicity Assays*

For all assays:

1. NP suspensions
2. Triton-X 100, 0.1% in cell culture medium
3. Respective cell culture medium

5.2.4.1 Alamar Blue

1. Fluorescent plate reader
2. CellTiter-Blue® (Promega), Alamar Blue® (Invitrogen) or similar, 10% in fresh culture medium
3. Opaque-walled tissue culture plates compatible with fluorimeter (clear or solid bottom)

5.2.4.2 Lactate dehydrogenase

1. Absorbance microplate reader
2. Lactate dehydrogenase (LDH) detection kit (Roche)
3. 96-well tissue culture plates compatible with photometer (clear bottom)

5.2.5 *Reduced Glutathione Quantification for Analysis of Oxidative Stress*

1. Fluorescent plate reader (Ex350/Em420)
2. Plate reader (absorption, 595 nm)

3. PBS
4. Redox quenching buffer (RQB): 20 mM HCl, 2.15 mM EDTA, 10 mM ascorbic acid
5. 5% trichloroacetic acid (TCA) in RQB
6. Potassium phosphate buffer 0.1 M (pH 6.9)
7. Potassium phosphate buffer 1 M (pH 7.0)
8. 5 mg/mL *o*-phthalaldehyde in methanol
9. 7.5 mM *n*-ethylmaleimide in RQB
10. Reduced glutathione (GSH) and oxidized GSH dimer (GSSG)
11. Coomassie Brilliant Blue G-250
12. 95% ethanol
13. 85% (w/v) phosphoric acid
14. Bovine serum albumin (BSA)
15. Black 96-well plates
16. Whatman #1 filter paper and funnel

5.2.6 *Expression and Quantification of Inflammatory Markers*

5.2.6.1 PCR

1. Real-time polymerase chain reaction (PCR) cycler (Biorad CFX96 or similar)
2. QuantiTect® Reverse Transcription Kit (Qiagen)
3. QuantiTect® Probe PCR Kit (Qiagen)
4. Template ribonucleic acid (RNA)
5. Primers and probes
6. Standard plasmids for quantification
7. 1.5 mL and 2 mL Eppendorf vessels (DNase- and RNase-free)
8. 96-well PCR plates (DNase- and RNase-free)
9. Ice for cooling

5.2.6.2 FACSArray

1. BD FACSArray
2. Plate shaker
3. Soluble mediator CBA flex sets, for example, interleukin-8 (IL-8), tumor necrosis factor (TNF) (BD)
4. Human soluble protein buffer master kit—CBA flex set (BD)

5.2.7 *Analysis of Cellular Functionality*

5.2.7.1 TEER (Caco-2)

1. Epithelial voltohmmeter (EVOM) (World Precision Instruments)
2. Chopstick electrodes (World Precision Instruments)
3. Heating plate
4. 12-well plates with Transwell inserts of 0.4 μm pore diameter (Corning)

5.2.7.2 Von Willebrand factor (HUVEC)

1. Fluorescent or confocal microscope
2. Image processing software for quantification, if necessary
3. 70% isopropanol or ethanol
4. Triton X-100, 0.1% in PBS as a negative control
5. PBS
6. Anti-hvWF antibody (e.g., Dako, monoclonal mouse anti-hvWF clone F8/86) dil 1:50 in 2% BSA solution in PBS
7. Secondary antibody, fluorescent-coupled (e.g., Molecular Probe, Alexa Fluor 568) dil 1:50 in 2% BSA solution in PBS
8. Nuclei-labeling solution: 4′,6-diamidino-2-phenylindole-dihydrochloride (DAPI) (1:15,000) diluted in 2% BSA solution in PBS
9. Optical plate with thin bottom

5.2.7.3 Albumin release (C3A)

1. Microtiter plate reader (absorbance, 450 nm)
2. Human albumin enzyme-linked immunosorbent assay (ELISA) quantitation set (Bethyl Laboratories, E80-129 or equivalent
3. Positive control for albumin release: medium containing 20% FCS
4. Negative control for albumin release: FCS-free medium
5. Coating buffer: 50 mM carbonate-bicarbonate, pH 9.6
6. Wash solution: 50 mM Tris, 140 mM NaCl, 0.05% Tween 20, pH 8.0
7. Blocking solution: 50 mM Tris, 0.14 M NaCl, 1% BSA, pH 8.0

Table 5.1 Numbers of cells plated on differently sized cell culture supports

Plate	96-well plate	24-well plate	12-well plate
HUVEC	6×10^3 cells	2.5×10^4 cells	4×10^5 cells
Caco-2	2×10^4 cells	6×10^4 cells	1.2×10^5 cells
C3A	1×10^5 cells	5×10^5 cells	1×10^6 cells

8. Sample diluent: 50 mM Tris, 0.14 M NaCl, 1% BSA, 0.05% Tween 20
9. Enzyme substrate for horseradish peroxidase (HRP): 1,3,5-trimethylbenzene (TMB)
10. Stop solution: 0.18 M H_2SO_4
11. 96-well plates

5.3 Methods

5.3.1 *Cell Culture and General Maintenance of Cells*

Table 5.1 gives an overview of cell numbers to be plated on differently sized surfaces. HUVECs are grown for one day before reaching confluence and use in experiments, Caco-2 are cultured and differentiated for 10 days in cell culture plates and 21 days in transwell plates, and C3A cells are seeded one day before reaching confluence and used in experiments.

5.3.1.1 HUVEC

1. Put 5 mL of 1% gelatin solution in a T75 flask and incubate for at least 5 min at 37°C.
2. Confluent cells are trypsinized using trypsin/EDTA and split in a 1:3 ratio in order to reach confluence after 48–72 h in culture. Frequent change of medium (once every 1–2 days) is recommended.

5.3.1.2 Caco-2

1. The cells are grown in a T75 flask and passaged every seven days.
2. Trypsinise confluent cells using trypsin/EDTA (10 min, 37°C, 5% CO_2) and add a fresh culture medium.

3. Cells are centrifuged for 4 min at 270 g and resuspended in a fresh cell culture medium.
4. Seed 0.5×10^6 cells into a new T75 flask. The medium is changed every second day.

5.3.1.3 C3A

1. Cells are grown in T75 flasks with weekly passaging and medium change every two to three days.
2. Trypsinise confluent cells using trypsin/EDTA (15 min, 37°C and 5% CO_2) and add a fresh culture medium.
3. Split cells in a 1:4 ratio for standard passaging. For exposures, prepare cell culture surfaces by covering in 0.1 mg/mL of collagen in sterile water, incubation at 37°C for 0.5 h to 2 h, removing the liquid and rinsing with sterile PBS, and seed cell numbers specified in Table 5.1, 24 h prior to exposures.

5.3.2 *Particle Preparation and Cell Exposures*

1. Weigh or pipette the required amount of NPs.
2. Suspend in sterile water containing 2% FCS at a maximum particle concentration of 1 mg/mL and sonicate for 15 min.
3. Dilute stock to required concentrations using the cell culture medium.
4. Remove the medium from cultured cells and apply particle suspensions immediately.

5.3.3 *Particle Characterization by DLS*

1. Weigh out or pipette the required amount of NPs.
2. Suspend in sterile water containing 2% FCS at a maximum particle concentration of 1mg/mL and sonicate for 15 min.
3. Dilute stock to required concentrations using the cell culture medium.
4. Measure the solutions in the Zetasizer:

    - Close the lid and turn on the instrument and wait for 30 min for the lasers to heat up and the laser signal to stabilize.
    - Start the Zetasizer software.

- Clean the cuvette and add 1 mL of the prepared sample to the cuvette.
- Open a measurement file and follow the on-screen instructions.
- Insert the cuvette into the instrument, wait for the temperature to stabilize, and start the measurement.

The result of the measurement is an intensity-weight size distribution. Each fraction is given a z-average (size value), a polydispersity index (PdI), and an intensity percentage. A histogram of size distribution is provided.

5. Measure the solutions in the Nanosight:
   - Samples are diluted with the respective medium to a final concentration of $\sim 10^8$ particles per mL.
   - 500 μL of the sample is injected into the sample chamber using a 1 mL or a 2 mL syringe. The laser is turned on, and the correct viewing area is readily located and viewed on the screen.
   - A short video (<2 min) is recorded and analyzed, from which the particle size distribution and concentration are reported.

5.3.4 *Toxicity Assays*

5.3.4.1 Alamar Blue

1. Plate cells and grow them to confluence according to the protocol and treat with NPs suspended in the respective cell culture medium. As a control for 100% cytotoxicity, cells are incubated with the cell culture medium containing 0.1% Triton X100. As a negative control, cells are incubated with the cell culture medium only.
2. Following the incubation, equilibrate the CellTiter-Blue® reagent to room temperature prior to use while protecting from direct light [23].
3. Remove the culture medium from cells, perform the washing step, if required, and add a fresh medium. The removed medium can be used for the LDH assay (Section 5.3.4.2).
4. In dark conditions add 10% of CellTiter-Blue® reagent (i.e., in a 24-well plate with 400 μL of medium; add 40 μL of the reagent)

and incubate for the required time, usually from 1–4 h, at 37°C (see Note 1).

5. A "time 0" fluorescence measurement can be taken (see Note 1).
6. Read the fluorescence after the established time in the plate reader at 579Ex/584Em.
7. Calculate the viability as fluorescent units per minute or percentage with respect to the control (untreated) sample value.

5.3.4.2 Lactate dehydrogenase

1. Plate cells in a 96-well plate, grow them to confluence according to the protocol, and treat them with NPs suspended in the respective cell culture medium. As a positive control (high control) cells are incubated with the cell culture medium with 0.1% Triton X100; as a negative control (low control) cells are incubated with the cell culture medium (see Note 2).
2. After incubation, transfer 100 μL of the supernatant to a new 96-well plate. The medium can be stored at 4°C for up to three days.
3. Prepare the LDH reagent from the LDH Kit from Roche: reconstitute the lyophilized powder in bottle 1 with 1 mL of double dist. water for 10 min and mix thoroughly. The dye solution in bottle 2 is ready to use. For 100 samples add 250 μL of bottle 1 to 11.25 mL of bottle 2.
4. Add 100 μL of the LDH reagent to every supernatant in the 96-well plate.
5. Incubate the plate for 3 min at room temperature in the dark.
6. Measure the absorbance in the microplate at 492 nm.
7. Calculate the cytotoxicity according to

$$\text{Cytotoxicity (\%)} = \frac{\text{Experimental value} - \text{Negative control}}{\text{Triton control} - \text{Negative control}} \times 100.$$

5.3.5 *Reduced Glutathione Quantification for Analysis of Oxidative Stress*

5.3.5.1 Preparation of extracts and protein pellets

1. Plate cells on 12-well plates, or in larger wells, and treat with NPs (see Section 5.3.2) for 2, 6, and 24 h.

2. Remove the medium and rinse cells with PBS. Transfer the plates on to ice.
3. Scrape cells into 1 mL of ice-cold PBS and transfer into an Eppendorf tube. Centrifuge (1,500 g, 4°C).
4. Remove supernatants and resuspend pellets in 100 μL of ice-cold TCA/RQB. Incubate on ice for 15 min and centrifuge (15,000 g, 4°C).
5. Transfer supernatants into a new tube and use directly for measurements or store at −80°C (see Note 3). Save pellets for a protein assay, either immediately or after storage at −80°C.

5.3.5.2 GSH assay

1. For GSH measurements [24], prepare a standard curve with serial dilutions of 100 to 3.125 μM of reduced GHS in TCA/RQB and use 100 μM of GSSG as a control for the specificity of the reaction. Use TCA/RQB as a blank.
2. In a black 96-well plate, pipette 19 μL of TCA/RQB into every well.
3. Add 10 μL of the standard or cell lysate sample. For the initial experiments, examine the potential of particles in the sample to interact with the assay (see Note 3).
4. Add 48 μL of 1 M potassium phosphate buffer (pH 7) and incubate for 5 min at room temperature.
5. Add 200 μL of 0.1 M potassium phosphate buffer (pH 6.9).
6. Add 29 μL of 5 mg/mL *o*-phthalaldehyde in methanol.
7. Incubate in the dark for 30 min.
8. Read the plate (365/430 nm)

5.3.5.3 Protein assay (Bradford; 25)

1. Prepare Bradford reagent by suspending 50 mg of Coomassie Brilliant Blue in 95% ethanol, adding 50 mL of 85% (w/v) phosphoric acid, and diluting the reagent to 500 mL.
2. Filter the reagent through Whatman #1 paper.
3. Prepare a standard curve using BSA of 1,000 to 0 μg/mL of protein in serial dilution.
4. Resuspend protein pellets in 100 μL of water, vortex.
5. Add 10 μL of samples or standards in triplicate into a 96-well plate.

6. Add 200 μL of Bradford reagent to the wells.
7. Read at 595 nm.
8. Normalize the GSH concentration to milligrams of protein.

5.3.6 *Expression and Quantification of Inflammatory Markers*

5.3.6.1 PCR

5.3.6.1.1 Reverse transcription reaction (see Note 4)

1. Isolate RNA according to standard methods and store at −80°C until the RT reaction takes place.
2. Thaw the template RNA on ice. Thaw all other reagents at room temperature (15–25°C) and store on ice.
3. Prepare the genomic DNA elimination reaction on ice according to the manufacturer's instructions (14 μL per reaction); use the 0.2 mL Eppendorf vessel. Multiple RT reactions should be performed if more than 1 μg total RNA is used.
4. Incubate for 2 min at 42°C and then place immediately on ice. Do not exceed an incubation time of 10 min.
5. Prepare the RT master mix on ice:

Volume/rxn	
Quantiscript reverse transcriptase	1 μL
Quantiscript RT buffer, 5x	4 μL
RT primer mix	1 μL
Total volume	6 μL

5. Add 6 μL of the master mix to each tube of template RNA from step 4, mix, and store in ice.
6. Incubate for 15 min at 42°C. The incubation steps (steps 6 and 7) can be conducted with a PCR machine.
7. Incubate for 3 min at 95°C to inactivate Quantiscript reverse transcriptase.
8. Store at −20°C until use or proceed directly with real-time PCR.

5.3.6.1.2 Semiquantitative real-time PCR using dual-labelled Taqman probes

1. Thaw the 2× QuantiTect Probe PCR master mix, template DNA, primer and probe solutions, and distilled water. Store on ice.
2. Prepare a master mix for triplicate measurement, including the standard plasmids (at least three different concentrations) and no template control (RNase-free water).

Component	Volume/ rxn	Final concentration
2× QuantiTect Probe PCR master mix	6.25 μL	1×
Primer A (50 μM)	0.1 μL	0.4 μM
Primer B (50 μM)	0.1 μL	0.4 μM
Probe (12.5 μM)	0.1 μL	0.1-0.2 μM
RNase-free water	4.45 μL	
Total volume	11 μL	

3. Mix the master mix thoroughly and dispense appropriate volumes into PCR tubes.
4. Add each 1.5 μL template DNA from the RT step to the individual PCR tubes. The volume of complementary DNA (cDNA) added should not exceed 10% of the final PCR volume.
5. Program the real-time cycler.

Step	Time	Temp.	Additional comments
PCR initial activation	15 min	95°C	HotStarTaq DNA polymerase activation step
Thermal cycling:			
Denaturation	10–15 s	94°C	
Annealing/extension	30–60 s	60°C	Combined annealing/ extension step
Number of cycles	35–45		Number of cycles dependent on the amount of template DNA and expression level of the target gene

Detection channel: Choose according to the fluorescence dye selected.

6. Place the PCR tubes in the real-time cycler and start the PCR program.
7. Analysis of the real-time PCR generally can be carried out automatically in the rtPCR cycler software by clicking the appropriate detection channel on the analysis window. The threshold will be defined for the best-fit standard curve, and the cDNA amount will be calculated automatically. Otherwise the threshold can also be defined manually by clicking and dragging the threshold line on the graph. The obtained cDNA amounts are to be calculated by using these equations:

$$\text{Relative mRNA concentration} = \frac{\text{Measured concentration}}{\text{Measured } \beta\text{-action concentration}}$$

$$\text{Fold induction} = \frac{\text{Relative mRNA concentration (sample)}}{\text{Relative mRNA concentration (negative control)}}$$

5.3.6.2 FACSArray

The BD FACSArray can analyze multiple cytokines in a multiplex experiment using fluorescent beads. Alternatively, standard ELISA assays for single cytokines, or other multiplex instruments, can be used.

1. Set up the assay protocol according to the manufacturer's instructions.
2. Prewet a round-bottom 96-well plate with a wash buffer and remove the wash buffer.
3. Prepare capture beads for analytes and standards according to the instructions. Add 50 μL of the capture beads to each well.
4. Pipette 25–50 μL of the cell culture supernatant (fresh or stored at −80°C) or standards into the wells.
5. Mix the plate for 5 min on a shaker (500 rpm) and incubate (1 h, room temperature).
6. Prepare a PE detection reagent for the analytes to be measured. Add 50 μL to each well.
7. Mix the plate for 5 min on a shaker (500 rpm) and incubate (2 h, room temperature).

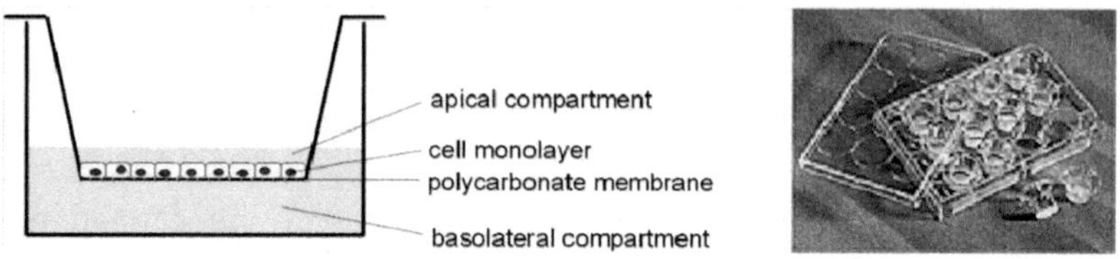

Figure 5.2 Transwell plate.

8. Remove the liquid from the wells and resuspend the beads by adding 150 μL of the wash buffer and shaking for 5 min (500 rpm).
9. Analyze samples on the BD FACSArray according to the manual.

5.3.7 *Analysis of Cellular Functionality*

5.3.7.1 TEER (Caco-2; Note 5)

1. Seed Caco-2 cells on transwell filter inserts (Fig. 5.2) at a seeding density of 60,000 cells/cm^2.
2. The electrode of the EVOM is stored in 0.1% KCl. Before the measurement, the electrode is sterilized with 70% ethanol for 15 min.
3. The electrode tip is then air-dried under the laminar flow hood.
4. Adjust the EVOM to "Resistance" measuring mode.
5. Cells are removed from the incubator and placed on a prewarmed heating plate (37°C).
6. The transepithelial electrical resistance (TEER) is measured in every well by placing the longer leg of the chopstick electrode at the bottom of the well and the shorter leg just above the cells in the cell culture insert (Fig. 5.3). Press the "Measure" button and wait 5 to 10 seconds for the measured resistance value to stabilize.
7. The TEER of blank Transwell filter inserts is measured and subtracted as background resistance from all values.
8. The TEER value can be used to monitor the formation of tight, functional epithelial cell layers, or fully differentiated Caco-2 cells as a measure of cell viability and functionality. TEER is normalized to the growth area of cells. Fully differentiated Caco-2

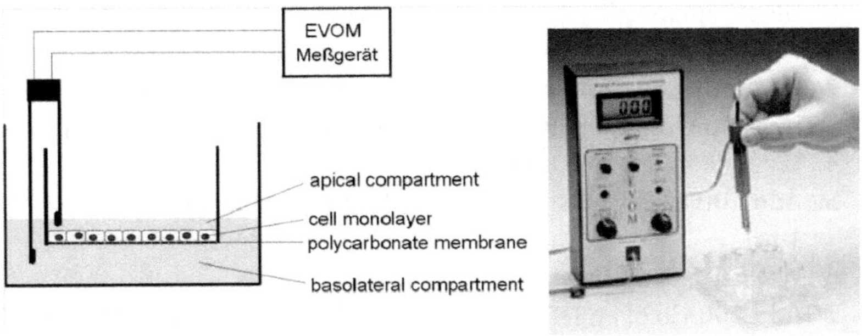

Figure 5.3 TEER measurement in Transwell with EVOM.

monolayers reach TEER values between 350 Ω/cm^2 and 600 Ω/cm^2.

5.3.7.2 Von Willebrand factor (Huvec; 26)

1. Remove the medium from the plate and wash each well twice with prewarmed PBS.
2. Add enough 70% isopropanol to cover the cells and fix at RT for at least 10 min.
3. Wash twice with PBS.
4. Add an hvWF antibody diluted to 1:50 in 2% BSA solution and leave at RT for 1 h.
5. Rinse three times with PBS.
6. Add a secondary antibody diluted to 1:50 in 2% BSA solution and leave at RT for 30 min in the dark.
7. Rinse three times with PBS.
8. Add DAPI diluted 1:15000 in 2% BSA solution and leave at RT for 5 min in the dark.
9. Rinse three times with PBS.
10. Examine cells using a confocal or fluorescence microscope.

5.3.7.3 Albumin release (C3A)

1. Dilute a capture antibody 1:100 in a coating buffer, add 100 μL/well into a 96-well plate, and incubate at room temperature for 1 h, or at 4°C overnight.
2. Wash the plate five times.

3. Block unspecific binding sites with 200 μL of a blocking solution per well (30 min, RT). In the meantime, prepare sample dilutions (see Note 6) and standards (400 to 0 ng/mL albumin in sample diluents according to the manufacturer's manual).
4. Wash the plate five times.
5. Add 100 μL of the standard or diluted sample and incubate (RT, 1 h).
6. Wash the plate five times.
7. Add 100 μL of an HRP detection antibody (see Note 6; RT, 1 h).
8. Wash the plate five times.
9. Add 100 μL of the TMB substrate solution and develop in the dark (15 min, see Note 6).
10. Add 100 μL of a stop solution to each well and measure on a plate reader (450 nm).

5.4 Notes

1. **Alamar Blue**. The ratio of CellTiter-Blue® reagent: The cell culture volume and incubation time should be adjusted for optimal performance, depending on the cell type, the metabolic capacity, the concentration of cells, and the linear range desired. Another method to analyse Alamar Blue is to calculate the specific metabolic activity of the cells, subtracting the "time 0" fluorescence value from the endpoint value and dividing the difference by time:

$$\text{SMActivity} = (\text{Abs}_{2h30'} - \text{Abs}_{30'})/120'$$

2. **LDH assay**. When seeding cells in a 96-well plate, it's better not to seed cells at the edge of the plate, Two problems arise at the edge of the plate: I) The medium more easily evaporates, and II) the thermal gradients are greatest at the edge of the plate, combining to give an uneven distribution of the cells on the well bottoms and uneven cell growth, resulting in higher scattering of cell numbers and data [27]. During the measurement the absorption must be watched. It might be necessary to modify the incubation time with the LDH reagent if a different number of cells are seeded and high or low LDH amounts are released to avoid positive control values exceeding the detection range.

3. **GSH assay**. GSH is stable for up to four weeks at −80°C. To ensure correct quantification it is advisable to freeze an aliquot of a known standard or standard curve at the same time as freezing the samples to ensure that the samples have not decayed in storage. Addition of 4 μL of 7.5 mM NEM in RQB with samples containing NPs inhibits the reaction and can be used as a background control for interference of the particles with the assay, for example by particle fluorescence.
4. **PCR**. The protocol is for use with 10 pg–1 μg RNA. If using >1 μg RNA, scale up the reaction linearly. If the RT-PCR product is longer than 200 bp, or if RNAs have very high secondary structure, increase the time of incubation at 42°C up to a total of 30 min. The QuantiTect Probe PCR master mix provides an initial 4 mM Mg^{2+}. For a few targets, reactions may be improved using a Mg^{2+} concentration of up to 6 mM. ß-actin also acts as housekeeping gene and can be substituted by other housekeeping genes.
5. **TEER**. TEER is not a constant readout but varies slightly due to positioning of the electrodes in the well. It is very sensitive to handling and temperature. Thus be quick but gentle in your handling of the cell culture plate and put the cell culture plate on a heating plate at 37°C during the measurement.
6. **Albumin ELISA**. The buffer compositions are for the Bethyl E80-129 kits and may vary for other kits. We diluted C3A culture supernatants 1:40 prior to measuring; this may vary for other hepatocytes. We also recommend running a standard curve and one control sample at different dilutions and different concentrations of a secondary antibody before analyzing a complete plate.

Acknowledgments

We would like to acknowledge funding from the European Union FP7 project InLiveTox (NMP4-SL-2009-228625).

References

1. British Standards. (2007) Terminology for nanomaterials. PAS 136, Publicly available specification, London: British Standards Institute.

2. Buseck, P. R., Tsipurski, S. J., Hettich, R. (1992) Fullerenes from the geological environment. *Science* **257**, 215–217.
3. Maynard, A. D. (2006) Safe handling of nanotechnology. *Nature* **444**, 267–269.
4. Oberdörster, G., Oberdörster, E., Oberdörster, J. (2005) Nanotoxicology: an emerging discipline evolving from studies of ultrafine particles. *Environ Health Perspect* **113**, 823–839.
5. Chaloupka, K., Malam, Y., Seifalian, A. M. (2010) Nanosilver as a new generation of nanoproduct in biomedical applications. *Trends Biotechnol* **28**, 580–588.
6. Saber, A. T., Koponen, I. K., Jensen, K. A., Jacobsen, N. R., Mikkelsen, L., Moller, P., Loft, S., Vogel, U., Wallin, H. (2012) Inflammatory and genotoxic effects of sanding dust generated from nanoparticle-containing paints and lacquers. *Nanotoxicology*, **6**(7), 776–788.
7. Osmond, M. J., McCall, M. J. (2010) Zinc oxide nanoparticles in modern sunscreens: an analysis of potential exposure and hazard. *Nanotoxicology* **4**, 15–41.
8. Cai, Z., Wang, Y., Zhu, L. J., Liu, Z. Q. (2010) Nanocarriers: a general strategy for enhancement of oral bioavailability of poorly absorbed or pre-systematically metabolized drugs. *Curr Drug Metab* **11**, 197–207.
9. Choi, H. S., Frangioni, J. V. (2010) Nanoparticles for biomedical imaging: fundamentals of clinical translation. *Mol Imaging* **9**, 291–310.
10. Brown, D. M., Wilson, M. R., MacNee, W., Stone, V., Donaldson, K. (2001) Size-dependent proinflammatory effects of ultrafine polystyrene particles: a role for surface area and oxidative stress in the enhanced activity of ultrafines. *Toxicol Appl Pharmacol* **175**, 191–199.
11. Semmler-Behnke, M., Kreyling, W. G., Lipka, J., Fertsch, S., Wenk, A., Takenaka, S., Schmid, G., Brandau, W. (2008) Biodistribution of 1.4- and 18-nm gold particles in rats. *Small* **4**, 2108–2111.
12. Schleh, C., Semmler-Behnke, M., Lipka, J., Wenk, A., Hirn, S., Schöffler, M., Schmid, G. N., Simon, U., Kreyling, W. G. (2011) Size and surface charge of gold nanoparticles determine absorption across intestinal barriers and accumulation in secondary target organs after local administration. *Nanotoxicology* **6**(1), 36–46.
13. Kang, B., Mackey, M. A., El-Sayed, M. A. (2010) Nuclear targeting of gold nanoparticles in cancer cells induces DNA damage, causeing cytokinesis arrest and apoptosis. *J Am Chem Soc* **132**, 1517–1519.

14. Li, N., Sioutas, C., Cho, A., Schmitz, D., Misra, C., and Sempf, J. (2003) Ultrafine particlulate pollutants induce oxidative stress and mitochondrial damage. *Environ Health Perspect* **111**, 455–460.
15. The Royal Society and the Royal Academy of Engineers. (2004) Nanoscience and nanotechnologies: opportunities and uncertainties. RS Policy Document 19/04, London, UK: Royal Society.
16. Chapter 1 in this book.
17. Chapter 8 in this book.
18. Stone, V., Johnston, H., Schins, R. P. (2009) Development of in vitro systems for nanotoxicology: methodological considerations. *Crit Rev Toxicol* **39**, 613–626.
19. Brown, D. M., Dickson, C., Duncan, P., Al-Attili, F., Stone, V. (2010) Interaction between nanoparticles and cytokine proteins: impact on protein and particle functionality. *Nanotechnology* **21**, 215104 (Epub).
20. Ju-Nam, Y., Lead, J. R. (2008) Manufactured nanoparticles: an overview of their chemistry, interactions and potential environmental implications. *Sci Total Environ* **400**, 396–414.
21. Maruyama, Y. (1963) The human endothelial cell tissue culture. *Z Zellforsch Mikrosk Anat* **60**, 69–79.
22. Baudin, B., Bruneel, A., Bosselut, N., Vaubourdolle, M. (2007) A protocol for isolation and culture of human umbilical vein endothelial cells. *Nat Prot* **2**, 481–485.
23. Technical Bulletin CellTiter-Blue® Cell Viability Assay, Instructions for use of Products G8080, G8081 AND G8082. Promega.
24. Senft, A. P., Dalton, T. P., Shertzer, H. G. (2000) Determining glutathione and glutathione disulfide using the fluorescent probe *o*-phthalaldehyde. *Anal Biochem* **280**, 80–86.
25. Bradford, M. M. (1976) Rapid and sensitive method for the quantitation of microgram quantities of protein utilizing the principle of protein-dye binding. *Anal Biochem* **72**, 248–254.
26. Galbusera, M., Zoja, C., Donadelli, R., Paris, S., Morigi, M., Benigni, A., Figliuzzi, M., Remussi, G., Remuzzi, A. (1997) Fluid shear stress modulates von Willebrand factor release from human vascular endothelium. *Blood* **90**, 1558–1564.
27. Lundholt, B. K., Scudder, K. M., Pagliaro, L. (2003) A simple technique of reducing edge effect in cell-based assays. *J Biomol Screen* **8**, 566–570.

Chapter 6

Utilizing Nanosensor-Incorporated Scaffolds in the Development of a 3D Lung Model

Helen C. Harrington,[a,b] Felicity R. A. J. Rose,[c] Amir M. Ghaemmaghami,[a] and Jonathan W. Aylott[b]

[a] *Allergy Research Group, School of Molecular Medical Sciences, University of Nottingham, U.K.*
[b] *Laboratory of Biophysics and Surface Analysis, School of Pharmacy, University of Nottingham, U.K.*
[c] *Division of Drug Delivery and Tissue Engineering, Centre for Biomolecular Sciences, University of Nottingham, U.K.*
jon.aylott@nottingham.ac.uk

The quest to develop more physiologically relevant human cell culture models that can be used for predictive preclinical studies has led to a movement toward developing 3D multicell culture models. Culturing cells in 3D on appropriate scaffolds is thought to better mimic the in vivo microenvironment and increase cell–cell interactions, therefore being more relevant to studying the molecular events and cell–cell interactions involved in the pathogenesis of diseases such as those relating to the lung. To create effective 3D cultures with high cell viability throughout the scaffold, the culture conditions such as oxygen and pH need to be

Cellular in vitro Testing: Methods and Protocols
Edited by John W. Haycock, Arti Ahluwalia, and John M. Wilkinson

ISBN 978-981-4364-97-3 (Hardcover), 978-981-4364-98-0 (eBook)
www.panstanford.com

carefully controlled as gradients in analyte concentration can exist throughout the 3D construct. Here we describe the development of biocompatible scaffolds for construction of a human lung model and a sensing strategy involving incorporation of analyte-responsive nanosensors into electrospun poly(lactic-co-glycolic acid) (PLGA) fibers to create novel self-reporting scaffolds. The self-reporting scaffolds can be used as tools to determine analyte concentrations within a scaffold microenvironment. Being able to monitor microenvironment conditions in real time without damaging the engineered tissue presents enormous opportunities to fully understand and optimize the production of 3D model tissue constructs.

6.1 Introduction

Lung diseases refer to disorders of the lung, such as asthma and chronic obstructive pulmonary diseases (COPD). The World Health Organization (WHO) estimates that in 2004 there were 235 million people suffering from asthma. Despite enhanced patient care, the morbidity and mortality of patients with asthma have remained high, with one asthma-related death every 20 min worldwide. This is partly due to a lack of efficient therapeutic strategies and also because a large proportion of patients do not respond to treatments. There is a lack of predictive preclinical models of asthma, and new treatments that enter clinical trials frequently fail, possibly because preclinical animal studies are often limited in their physiological relevance to the human lung [1]. One way to circumvent such limitations is to develop biomimetic models of the human lung, where anatomical and functional properties of the human lung are simulated using human-derived cells grown in three-dimensions on relevant scaffolds.

The airway epithelium provides a primary barrier against the inhaled environment, and as such epithelial cells are commonly used for in vitro studies of lung disease and drug discovery or delivery studies. Respiratory epithelial cells cultured at the air–liquid interface (ALI) generally have morphological and functional characteristics of the in vivo airway epithelium as opposed to submerged cultures [2, 3]. As a result there are various commercially available permeable cell culture membranes enabling two-dimensional (2D) culture at the ALI and transepithelial effects to be studied. The response of epithelial cells to various environmental stimuli can be assessed by monitoring membrane integrity or the permeation of molecules across the paracellular space [4].

In addition to providing a physical barrier, the epithelium dynamically responds to environmental stimuli by interacting with other key cell types such as immune responsive cells or subepithelial cells for innate host defense [5–7]. Therefore a more relevant in vitro lung model would have a multicell composition of epithelial, airway smooth muscle, and immune responsive cells, in which the cells can interact to better mimic in vivo conditions.

Cocultures can be created by seeding cells onto the apical side of a cell culture insert whilst another cell type is seeded upon the basal side [8, 9]. A similar method of coculture has recently been used to demonstrate that secretory factors produced by the Calu-3 cells grown on a membrane insert can interact with peripheral blood mononuclear cells (PBMCs) that had been seeded into the well plate [10]. Membrane inserts provide a suitable platform to allow cell interaction via secretory factors; however, it is becoming apparent that cells can interact further by physical contact [11].

Tissue engineering approaches can be used to create more complex cell culture models that have a three-dimensional (3D) architecture, allowing greater cell–cell interactions and resulting in more physiologically relevant models. We have developed a multicell, multilayer approach using electrospun tissue-engineered scaffolds. Epithelial cells form the uppermost layer of the 3D construct, fibroblasts form the base, and immune responsive dendritic cells are incorporated beneath the epithelial layer. Each layer has been developed to accommodate the requirements of the inoculated cells. Epithelial and fibroblast cells are each cultured on a biocompatible electrospun polymer scaffold. The scaffold morphology resembles that of lung tissue and provides mechanical support for cell attachment and proliferation yet allows cell migration and interaction throughout the interstices of the 3D structure. Production of matrix proteins, growth factors, and cytokines from these cells and dendritic cells inoculated into a gel layer should facilitate the development of a metabolically active model lung. Furthermore the

3D model can be positioned so that the epithelial cells are cultured at the ALI, allowing differentiation of epithelial cells.

A practice that is popular in the production of component parts of the lung for transplantation studies is to culture epithelial cells on decellularized lung tissue, giving the advantage that natural extracellular matrix (ECM) proteins are present to assist in the repopulation of the lung matrix [12, 13]. However, for pathogenesis studies this approach has the limitations of batch-to-batch variability of tissue, and the availability of lung tissue is limited. Furthermore, decellularized tissue cannot be structurally modified nor can drug particles for controlled release be incorporated easily.

Scaffold carrier systems that release drugs for treatment purposes can be fabricated using synthetic material as the tissue matrix [14]. This concept can be adapted to incorporate sensing devices into the scaffold enabling in situ assessment of environmental conditions within a 3D cellular construct. Monitoring environmental conditions is generally performed by physically probing the construct, which can disturb cellular function. It is known that within 3D constructs oxygen gradients can exist with cells at the centre experiencing hypoxia, resulting in inhomogeneous tissue quality [15]. The use of perfusion bioreactors can prevent cell death but may not entirely eliminate 3D culture–associated oxygen gradients. Therefore continuous oxygen monitoring of 3D culture systems to ensure tissue quality throughout engineered constructs is recommended. We have incorporated optically responsive nanosensors into polymer fibers to produce a novel self-reporting scaffold capable of noninvasive in situ assessment of local environmental analyte concentrations, such as oxygen, pH, and calcium. Scaffolds that act as sensing devices by monitoring fluorescence activity have been prepared previously, where the sensing device was either the actual polymeric scaffold created by electrospinning or an analyte-responsive dye, which is incorporated into the polymer prior to scaffold formation [16, 17]. However, these sensing devices are/were prone to interference from other analytes and therefore may give erroneous optical outputs. The use of a ratiometric sensing device holds the potential to eliminate these possible adverse effects and provide a response specific to the analyte in question. Coupling the use of self-reporting scaffolds and perfusion bioreactors can in the future provide online assessment of a dynamic 3D culture.

6.2 Materials

All materials, unless stated otherwise, were obtained from Sigma (U.K.) and used as received.

6.2.1 *Nanosensor Preparation*

1. Use a Millipore® anodized filter membrane (0.02 μm) in a Buchner funnel for vacuum filtration of nanosensors.
2. Following preparation, store nanosensors in a glass vial at 4°C.

6.2.2 *Preparation of Oxygen-Responsive Sol-Gel Nanosensors*

1. Reagents used are ethanol, anhydrous dimethylformamide (DMF) and tetraethyl orthosilicate (TEOS), fluorophores Oregon Green Dextran, tris (4,7-diphenyl-1,10-phenanthroline) ruthenium(II) chloride ($Ru(dpp)_3{}^{2+}$) (GFS Chemicals, USA), and ammonium hydroxide 50% v/v aqueous solution (Alfa Aesar, U.K.) diluted to 30% v/v.

6.2.3 *Preparation of pH-Responsive Sol-Gel Nanosensors*

1. Reagents used are ethanol, anhydrous DMF, TEOS, 3-aminopropyltriethoxysilane (APTES), fluorophores 5-carboxyfluorescein and 6-carboxyfluorescein, succinimidyl ester (FAM-SE), 6-carboxytetramethylrhodamine succinimidyl ester (TAMRA-SE) (Invitrogen), and ammonium hydroxide 50% v/v aqueous solution (Alfa Aesar, U.K.) diluted to 30% v/v.

6.2.4 *Nanosensor Response to Analytes*

1. Nanosensors were suspended in deionized water obtained from an ELGA purification system, resistivity 18 Ω·cm (Maxima USF ELGA, U.K.).

2. Nanosensor solutions were contained in an optical cuvette suitable for fluorescence measurements.
3. A Varian Cary Eclipse fluorescence spectrophotometer was used. The nanosensor solutions were excited using a xenon flash lamp with excitation and emission slits set at 5 nm bandwidths.

6.2.5 *Nanosensor Response to Oxygen*

1. Oxygen gas and nitrogen gas (BOC) were mixed using a Signal Series 850 gas blender equipped with a temperature controller. The gas blender operates by producing a constant flow of gas by means of a pair of differential pressure regulators that control the flow of oxygen and the nitrogen diluent gas, taking into account temperature and specific gravity of the gas. The flow rates of both gases are controlled to give exact oxygen concentrations. The oxygen concentration exiting the gas blender was verified using an oxygen analyzer (GB300 Cambridge Sensotec).

6.2.6 *Nanosensor Response to pH*

1. Prepare Sørensen's phosphate buffer solutions ranging from pH 5.5 to pH 7.5 by mixing specific ratios of monobasic (0.2 M) and dibasic (0.2 M) stock solutions to form various sodium phosphate buffers.
2. Check the final pH using a pH meter (Jenway model 3510).
3. Make any minor adjustments to the pH using NaOH or HCl.

6.2.7 *SEM Analysis*

1. Place samples, nanosensors, PLGA scaffolds, or self-reporting scaffolds onto carbon-coated electron microscope stubs.
2. Sputter-coat samples with gold (5 min, Blazers SCD 030 gold sputter coater) under an argon atmosphere (BOC) prior to analysis.
3. Place samples under a scanning electron microscope (SEM) (Jeol JSM 60602v).

6.2.8 *Electrospinning PLGA Scaffolds*

1. PLGA (70:30 LA:GA ratio, 0.45 g) (SurModics Pharmaceuticals, USA) stored at –20°C, dissolved in dichloromethane (DCM) (HPLC grade, Fisher Scientific) to 20% wt/wt with pyridinium formate (PF) 1% wt/wt
2. An 18-gauge blunt-end needle, internal diameter of 0.8 mm (Becton Dickinson), and a syringe pump (Harvard PHD 2200 syringe pump)
3. High-voltage power supply (Glassman High voltage Supply Series EL)
4. A 25 × 35 cm aluminum sheet

6.2.9 *Electrospinning Self-Reporting PLGA Scaffolds*

1. Nanosensors (5 mg/mL of polymer solution) are required, in addition to the materials and equipment necessary for electrospinning PLGA scaffolds.

6.2.10 *Self Reporting Scaffold Analysis*

1. Confocal microscope (Leica TCS-SP) equipped with argon and krypton lasers, the laser emissions being 488 nm and 568 nm, respectively. Images were obtained using a 63×, 0.9 NA water immersion lens, with samples immersed in phosphate buffered saline (PBS) solution. Samples were contained in 35 mm cell culture plates (Beckton Dickenson Labware, U.K.).

6.2.11 *Calibration of the Oxygen-Responsive Scaffold*

1. The materials and instruments described for nanosensor response to oxygen were used.

6.2.12 *Calibration of the pH-Responsive Scaffold*

1. The materials and instruments described for nanosensor response to pH were used.

6.3 Methods

6.3.1 *Nanosensor Preparation*

The sol-gel process involves the evolution of inorganic networks through the formation of a colloidal suspension (sol) and gelation of the sol to form a network in a continuous liquid phase [18]. It is possible to vary the structure and properties of sol-gel products by controlling parameters that affect the rates of hydrolysis and condensation reactions, such as the pH, nature, and concentration of the catalyst and the H_2O/Si molar ratio (R) [19]. The Stöber process, whereby the reaction is catalyzed by ammonia in the presence of water and a low-molecular-weight alcohol and having an R value of 7–25, can be used to produce monodisperse spherical nanoparticles [20]. Including dyes for ratiometric pH and oxygen sensing in the reaction mixture can result in their incorporation into the nanoparticles. Further details of the Stöber process can be found in Note 1.

6.3.2 *Preparation of Oxygen-Responsive Sol-Gel Nanosensors*

1. Add ethanol (6 mL) to a round-bottomed flask.
2. Add the fluorophores $Ru(dpp)_3^{2+}$ (0.005 g) and Oregon Green Dextran (5 mg/mL, 50 μL) to the round-bottomed flask. $Ru(dpp)_3^{2+}$ and Oregon Green Dextran are sensing and reference dyes, respectively, and provide a ratiometric response to oxygen. Further details of the ratiometric response to an analyte are given in Note 2, and details of the response to oxygen are given in Note 3.
3. Add ammonium hydroxide (30% v/v, 4 mL) to the mixture.
4. Stir the mixture for 1 h.
5. Add TEOS (0.5 mL) drop-wise to the mixture and continue stirring for a further 2 h.
6. Collect the nanosensors by centrifugation (300 g, 20 min).
7. Wash the nanosensors with ethanol (10 mL) and deionized water (10 mL) and further centrifuge, followed by discarding the supernatant.

8. Repeat the washing process several times to remove any unreacted chemicals.
9. Collect the nanosensors by vacuum filtration using a Millipore® anodized filter membrane (0.02 μm) and dry under vacuum.
10. The fluorescence emission for oxygen-responsive scaffolds can be collected at 500–530 nm for Oregon Green Dextran and 600–630 nm for $Ru(dpp)_3{}^{2+}$ (Fig. 6.1A).

6.3.3 *Preparation of pH-Responsive Sol-Gel Nanosensors*

1. Dissolve FAM-SE (1.5 mg) (Invitrogen) in DMF in a round-bottomed flask; add to APTES (1.5 mL) and stir under a dry nitrogen atmosphere for 24 h in the dark. FAM is a fluorescein derivative and is pH responsive. Further details of the fluorescence of FAM are given in Note 4, and the use of FAM-SE is given in Note 5.
2. Dissolve TAMRA-SE (1.5 mg) (Invitrogen) in DMF in a round-bottomed flask; add to APTES (1.5 mL) and stir under a dry nitrogen atmosphere for 24 h in the dark. TAMRA is the reference dye for pH nanosensors, and the use of TAMRA-SE is detailed in Note 5.
3. Add both of the above dyes (250 μL) to ethanol (6 mL) and ammonium hydroxide (30% v/v, 4 mL) contained in a round-bottomed flask and stir for 1 h.
4. Add TEOS (0.5 mL) drop-wise to the mixture; stir for a further 2 h.
5. Collect the nanosensors by vacuum filtration using a Millipore® anodized filter membrane (0.02 μm) and dry under vacuum.
6. The fluorescence emission for pH-responsive scaffolds can be collected at 500–530 nm for FAM-SE and 558–580 nm for TAMRA-SE (Fig. 6.1C).

6.3.4 *Nanosensor Response to Oxygen*

1. Suspend nanosensors in deionized water (5 mg/mL) by methods of ultrasonication and vortexing.
2. Place the solution in an optical cuvette.

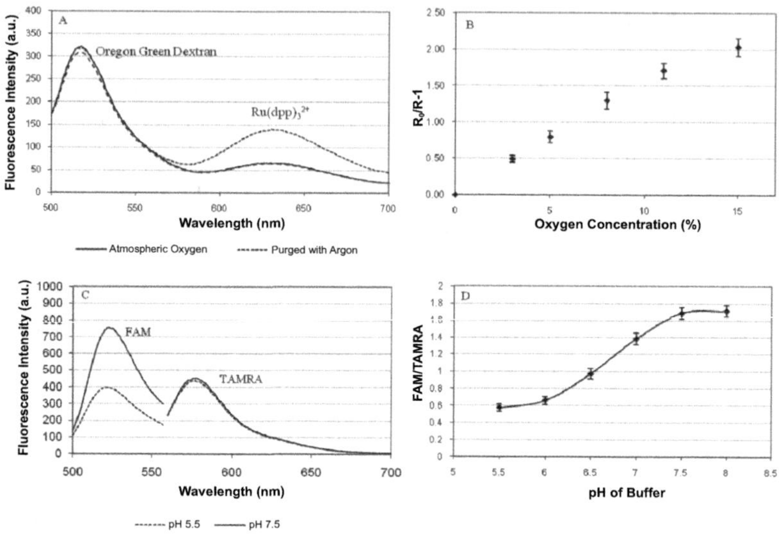

Figure 6.1 (A) Fluorescence emission spectra of oxygen-responsive sol-gel nanosensors, showing the response of the incorporated dyes to oxygen. (B) Stern–Volmer plot for oxygen-responsive sol-gel nanosensors over a biologically relevant oxygen concentration range. (C) Fluorescence emission spectra for pH-responsive sol-gel nanosensors at pH 5.5 and pH 7.5. (D) Calibration curve for pH-responsive sol-gel nanosensors, calculated using the ratio of the fluorescence intensities FAM and TAMRA over a biologically relevant pH range. Error bars represent standard error of the mean and $n = 3$.

3. To determine the response to a change in oxygen concentration, purge the solution with different concentrations of oxygen ranging from 0% to 21%.
4. Collect the emission spectra using a fluorescence spectrometer.
5. Input the data into the Stern–Volmer equation to create a calibration curve (Fig. 6.1B).

6.3.5 *Nanosensor Response to pH*

1. Suspend pH nanosensors in different pH buffers (5 mg/mL) by methods of ultrasonication and vortexing.
2. Place the first solution in an optical cuvette.
3. Collect the emission spectra using a fluorescence spectrometer.

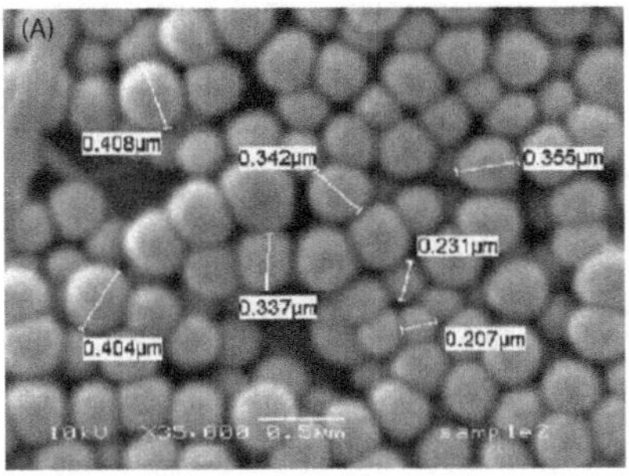

(B)

	Oxygen nanosensors	pH nanosensors
	nm	
Mean diameter	317	342
St dev	48	47
Max	440	470
Min	197	243

Figure 6.2 (A) Representative SEM micrograph of oxygen-responsive sol-gel nanosensors produced using the Stöber process. (B) Measurements for oxygen and pH nanosensors.

4. Change the solutions in the optical cuvette and continue monitoring the emission spectra.
5. Calculate the ratio of the emission maxima produced at each pH value to produce a calibration curve (Fig. 6.1D).

6.3.6 *Scanning Electron Microscopy of Nanosensors*

1. Place a sample of the nanosensors onto a carbon-coated electron microscope stub.
2. Sputter-coat with gold for 5 min under an argon atmosphere.
3. Observe under an SEM. During imaging adjust the working distance, voltage, and magnification to minimize electron charging (Fig. 6.2).

6.3.7 *Electrospinning PLGA Scaffolds*

1. Dissolve PLGA in DCM (20% w/w) with PF (1% w/w) by stirring for 24 h at room temperature.
2. Place the solution in a 10 mL syringe with an 18 G blunt fill needle and securely fit it to the syringe pump.
3. Distance the needle tip 20 cm away from the 20 × 15 cm aluminum collecting plate.
4. Attach the electrode of the high-voltage power supply to the tip of the syringe and the earth to the aluminum collecting plate.

5. Deliver the solution using a constant flow rate of 3.5 mL/min at 12 kV for 1.5 h.
6. Leave the scaffold in a fume hood for 24 h to allow the solvent residue to evaporate.

6.3.8 *Electrospinning Self-Reporting PLGA Scaffolds*

1. Self-reporting scaffolds were prepared as above but with the modification of adding nanosensors to the polymer solution (5 mg/mL). The nanosensors were suspended in the PLGA solution with the assistance of ultrasonication prior to placing in a 10 mL syringe.

6.3.9 *Scanning Electron Microscopy of the PLGA Scaffold and Self-Reporting Scaffold*

1. Place a sample of the PLGA scaffold or self-reporting scaffold onto a carbon-coated electron microscope stub and proceed as described for scanning electron microscopy of nanosensors (Fig. 6.3).

6.3.10 *PLGA Scaffold Analysis: Calibration of the Oxygen-Responsive Scaffold*

1. Place a sample of the oxygen-responsive scaffold in a 35 mm culture plate and immerse in PBS (2 mL).
2. Pass gas mixtures at different oxygen concentrations in the PBS. Allow an equilibrating period of 15 min before taking measurements.
3. Observe the fluorescence emission for oxygen-responsive scaffolds using confocal microscopy at 500–530 nm for Oregon Green Dextran and 600–630 nm for $Ru(dpp)_3^{2+}$ (Fig. 6.4A).

6.3.11 *Calibration the of pH-Responsive Scaffold*

1. Place a sample of the pH-responsive scaffold in a 35 mm culture plate and immerse in buffers of different pH (2 mL).

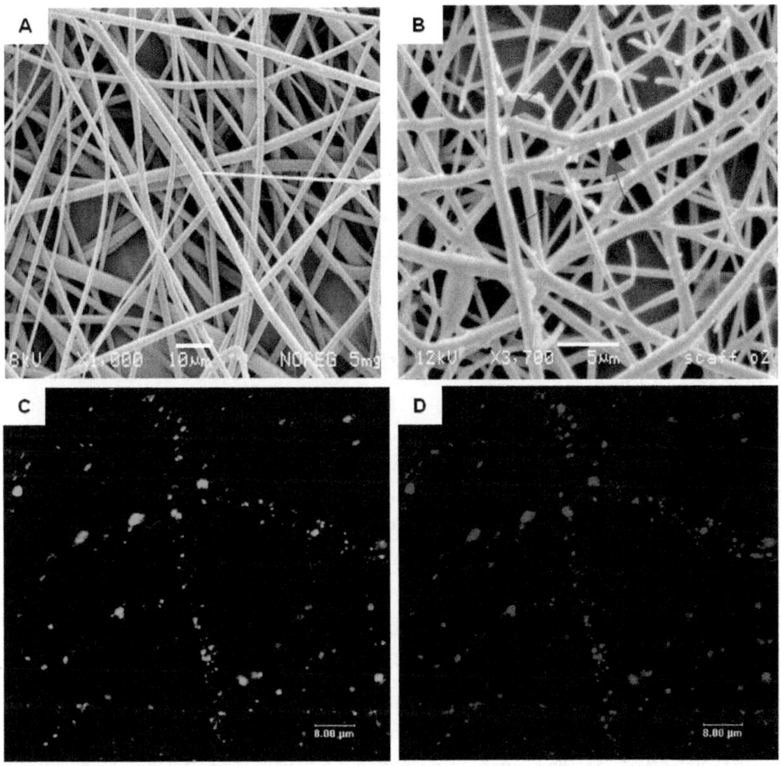

Figure 6.3 Representative SEM micrographs showing (A) control PLGA fibers (not incorporating nanosensors). Scale bar = 10 μm. (B) PLGA fibers electrospun with 5 mg/mL nanosensors, where the arrows indicate nanosensors associated with the fibers. Scale bar = 5 μm. Representative CLSM images of nanosensors incorporated into PLGA scaffolds to create self-reporting scaffolds. (C) FAM. (D) TAMRA. Scale bar = 8 μm. *Abbreviation*: CLSM, confocal laser scanning microscopy.

3. Observe the fluorescence emission for the pH-responsive scaffold intensity using confocal microscopy at 500–530 nm for FAM and 558–580 nm for TAMRA (Fig. 6.4B).

The calibration curves produced using spectrophotometry and CLSM are similar, although performed on different instrumentation. This indicates that the nanosensors require calibration upon each instrument used; however, this is to be expected, as the light source

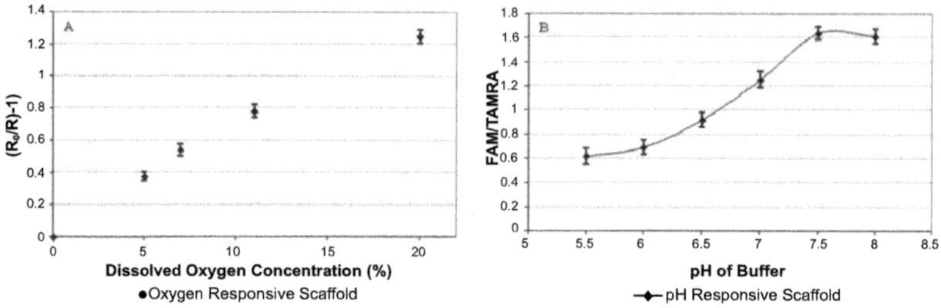

Figure 6.4 (A) Response of oxygen-sensitive sol-gel nanosensors incorporated into a PLGA scaffold. (B) Response of pH-sensitive sol-gel nanosensors incorporated into a PLGA scaffold. Error bars represent standard error of the mean and $n = 3$.

intensities are different for each instrument. In addition TAMRA has to be excited using a nonoptimal wavelength on the CLSM using the argon laser (488 nm) because the krypton laser (568 nm) is obtrusive to the emission of the dye and can give misleading results.

6.4 Notes

1. The Stöber process proceeds via hydrolysis of TEOS in a solution of ethanol, water, and ammonia. Ethanol is present because water and alkoxides are immiscible and an alcohol acts as a mutual solvent. During hydrolysis the alkoxide groups (OR) are substituted with hydroxyl groups (OH). Complete hydrolysis is achieved through use of a basic catalyst such as ammonia. It is likely that water dissociates under basic conditions to produce hydroxyl anions. Subsequent condensation reactions of the silanol groups (Si-OH) produce siloxane bonds (Si-O-Si) plus the by-products water or alcohol. To produce spherical particles the hydrolysis reaction should be complete before condensation begins. Changing conditions such as pH, H_2O/Si molar ratio (R), or addition of a catalyst can force the completion of hydrolysis. Increasing the R value promotes the hydrolysis reaction because there will be more hydroxyl anions present at the start to react with the silicon alkoxide. Additionally higher R values cause more complete hydrolysis of monomers before significant

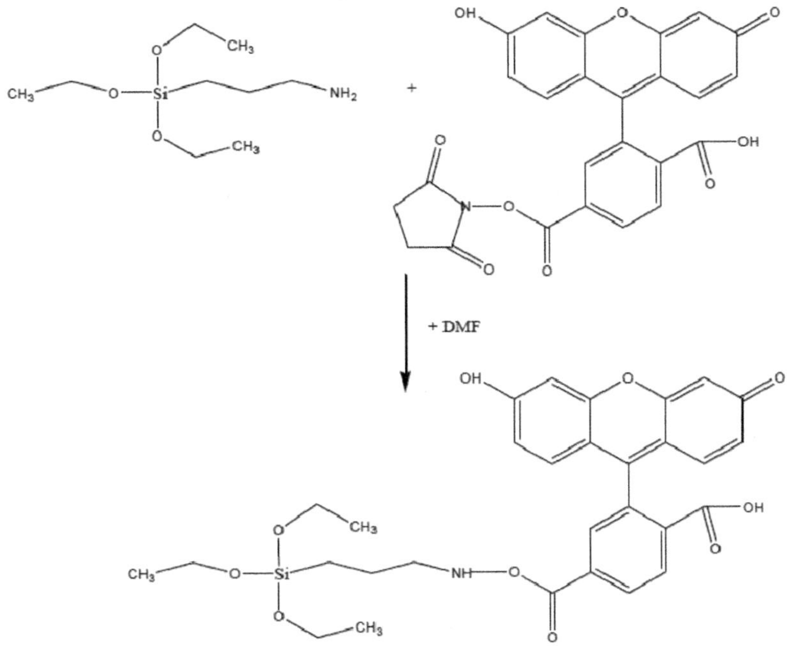

Figure 6.5 Stöber process performed using basic conditions. Water dissociates to produce hydroxyl anions in a rapid first step. The hydroxyl anions then attack the silicon atom in a condensation reaction. Further condensation produces siloxane bonds.

condensation occurs; this is a crucial requirement to produce spherical nanoparticles. Differing extents of monomer hydrolysis should affect the relative rates of the alcohol or water, producing condensation reactions, and since water is the by-product of the condensation reaction, large values of *R* promote siloxane bond hydrolysis; hence the condensation reaction is rapid. As the number of siloxane bonds increases the individual molecules are bridged and the polymer chains increase in length. During the two reaction steps of hydrolysis and condensation polymer solubility decreases until the chain precipitates out of solution as nanosized silica particles. The reaction scheme for the Stöber process is shown in Fig. 6.5.

2. The incorporation of a reference dye that acts as an internal standard within the sensing nanoparticles can eliminate delete-

rious effects such as variations in temperature and fluctuations in excitation source intensity. Incorporating both reference and sensing dyes into the same nanoparticle ensures that they are both reporting from the same environment. The nanoparticle polymer matrix is permeable to the analyte in question, allowing the analyte to interact with the entrapped dyes. Nanosensor response can be determined from the ratio of the maximum fluorescence intensity of the sensing dye to that of the reference dye $(I_{\text{sensing dye}}/I_{\text{reference dye}})$.

3. Optical oxygen sensors operate on the principle of oxygen molecules quenching the fluorescence of fluorophores through a process known as dynamic quenching; this approach benefits from fast response times and does not consume oxygen. Dynamic quenching occurs when the excited fluorophore collides with the quenching species, in this case a molecule of oxygen, resulting in nonradiative transitions to the ground state without fluorescence emission. Since the period of time that the fluorophores spend in the excited state is reduced, the overall lifetime of the fluorophore is also reduced. Thus, dynamic quenching reduces not only the fluorescence emission intensity but also the fluorophore emission lifetime. The degree of fluorescence quenching relates to the frequency of collisions with the quenching molecule and therefore to the concentration of the quencher within a sample. The quenching effect is therefore predictable and can be described by the Stern–Volmer relationship, which relates fluorescence emission intensity to the concentration of the quenching species: $I_0/I = 1 + \mathrm{K_{SV}}[Q]$, where I_0 and I are the fluorescence emission intensities observed in the absence and presence, respectively, of the quencher. The quencher concentration is represented by $[Q]$, and the Stern–Volmer quenching constant by $\mathrm{K_{SV}}$. A plot of $(I_0/I) - 1$ versus $[Q]$ should yield a straight line with a slope equal to $\mathrm{K_{SV}}$. A linear plot is desirable as it is indicative of the fluorophores reporting from the same environment, whereas a nonlinear Stern–Volmer plot is an indication that some fluorophores are less accessible to the quencher molecule than others.
4. The equilibrium distribution of the fluorophore fluorescein and its derivatives, in this case FAM-SE, can be monitored

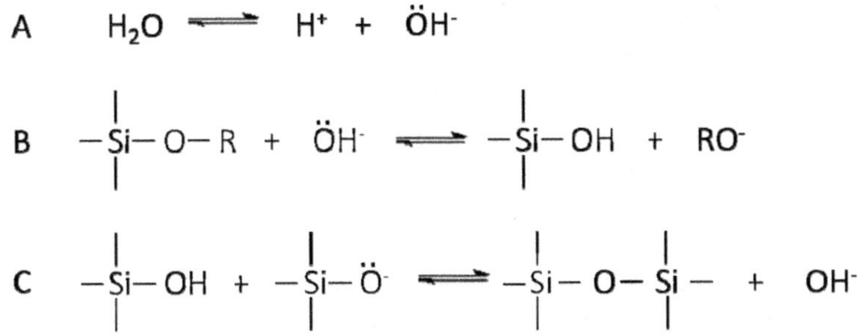

Figure 6.6 Covalent attachment of FAM-SE to APTES.

using fluorescence measurements to determine proton activity. Fluorescein fluoresces only when present in its monoanion and dianion forms, allowing local pH measurements to be determined by measuring the change in fluorescence intensity as equilibrium is established between these forms. It is possible to create a calibration curve as a result of the fluorescence emission produced by fluorescein; however, it can only be used for monitoring pH in the range pH 5 to pH 9 because the monoanion and dianion forms only exist in this pH range. This is because the phenol and carboxylic acid functional groups of fluorescein are almost completely ionized in aqueous solutions above pH 9 and become independent of pH. As the pH becomes more acidic the dianion protonates the monoanion and the fluorescence decreases until the neutral nonfluorescent species is produced at pH 5.

5. The succinimidyl ester versions of FAM (FAM-SE) and TAMRA (TAMRA-SE) have been used in this work because the succinimidyl ester group can be covalently linked via carboxyamide bonds to an amine-functionalized polymer matrix such as APTES (Fig. 6.6). The FAM/TAMRA-APTES construct can then be incorporated with TEOS in the sol-gel process.

References

1. Bragonzi, A. (2010) Murine models of acute and chronic lung infection with cystic fibrosis pathogens. *Int J Med Microbiol* **300**, 584–593.

2. Ross, A. J., Dailey, L. A., Brighton, L. E., Devlin, R. B. (2007) Transcriptional profiling of mucociliary differentiation in human airway epithelial cells. *Am J Resp Cell Mol Biol* **37**, 169–185.
3. Foster, K. A., Avery, M. L., Yazdanian, M., Audus, K. L. (2000) Characterization of the Calu-3 cell line as a tool to screen pulmonary drug delivery. *Int J Pharm* **208**, 1–11.
4. Miyoshi, J., Takai, Y. (2005) Molecular perspective on tight-junction assembly and epithelial polarity. *Adv Drug Delivery Rev* **57**, 815–855.
5. Knight, D. R. (2001) Epithelium-fibroblast interactions in response to airway inflammation. *Immunol Cell Biol* **79**, 160–164.
6. Kauffman, H. F. (2003) Interaction of environmental allergens with airway epithelium as a key component of asthma. *Curr Allergy Asthma Rep* **3**, 101–108.
7. Gershwin, L. J. (2007) Effects of allergenic extracts on airway epithelium. *Curr Allergy Asthma Rep* **7**, 357–362.
8. Rothen-Rutishauser, B. M., Kiama, S. G., Gehr, P. (2005) A three-dimensional cellular model of the human respiratory tract to study the interaction with particles. *Am J Resp Cell Mol Biol* **32**, 281–289.
9. Diabate, S., Mulhopt, S., Paur, H. R., Krug, H. F. (2008) The response of a co-culture lung model to fine and ultrafine particles of incinerator fly ash at the air-liquid interface. *ATLA, Altern Lab Anim* **36**, 285–298.
10. Torvinen, M., Campwala, H., Kilty, I. (2007) The role of IFN-gamma in regulation of IFN-gamma-inducible protein 10 (IP-10) expression in lung epithelial cell and peripheral blood mononuclear cell co-cultures. *Resp Res* **8**, 80.
11. Blank, F., Wehrli, M., Lehmann, A, et al. (2011) Macrophages and dendritic cells express tight junction proteins and exchange particles in an in vitro model of the human airway wall. *Immunobiology* **216**, 86–95.
12. Badylak, S. F., Freytes, D. O., Gilbert, T. W. (2009) Extracellular matrix as a biological scaffold material: structure and function. *Acta Biomater* **5**, 1–13.
13. Petersen, T. H., Calle, E. A., Zhao, L. P., et al. (2010) Tissue-engineered lungs for in vivo implantation. *Science* **329**, 538–541.
14. Saltzman, W. M., Olbricht, W. L. (2002) Building drug delivery into tissue engineering. *Nat Rev Drug Discovery* **1**, 177–186.
15. Volkmer, E., Drosse, I., Otto, S., et al. (2008) Hypoxia in static and dynamic 3D culture systems for tissue engineering of bone. *Tissue Eng Part A* **14**, 1331–1340.

16. Wang, X. Y., Drew, C., Lee, S. H., Senecal, K. J., Kumar, J., Sarnuelson, L. A. (2002) Electrospun nanofibrous membranes for highly sensitive optical sensors. *Nano Lett* **2**, 1273–1275.
17. Yang, Y., Yiu, H. H. P., El Haj, A. J. (2005) On-line fluorescent monitoring of the degradation of polymeric scaffolds for tissue engineering. *Analyst* **130**, 1502–1506.
18. Hench, L. L., West, J. K. (1990) The sol-gel process. *Chem Rev* **90**, 33–72.
19. Ahn, B. Y., Il, Seok, S., Baek, I. C. (2008) Sol-gel micro encapsulation of hydrophilic active compounds from the modified silicon alkoxides: the control of pore and particle size. *Mat Sci Eng C-Biomimetic Supramol Syst* **28**, 1183–1188.
20. Stober, W., Fink, A., Bohn, E. (1968) Controlled growth of monodisperse silica spheres in micron size range. *J Colloid Int Sci* **26**(10), 62.

Chapter 7

Development of a Paracrine-Signaling Model for Detecting Irritancy Using Human Keratinocytes and Fibroblasts

Jirapat Chunthapong,[a] Sheila MacNeil,[b] and John W. Haycock[b]
[a] *Department of Biology, Faculty of Science, Khon Kaen University, Khon Kaen 40002, Thailand*
[b] *Department of Materials Science and Engineering, Kroto Research Institute, Sheffield University, Sheffield, U.K.*
j.w.haycock@sheffield.ac.uk

Tissue-engineered materials for clinical purposes have led to the development of in vitro models as alternatives to animal testing. The development of this work is to understand and apply the basic principles of paracrine interactions arising between human keratinocytes and fibroblasts for detecting an irritant-induced inflammatory reaction for future testing purposes. We have used two irritants, sodium dodecyl sulphate (SDS) and potassium diformate (Formi®) at subtoxic concentrations and studied interleukin-1 alpha (IL-1α) release from human keratinocytes as the primary response to an irritant. Fibroblasts were cultured separately and used to link the primary cytokine response to a series of secondary responses using a keratinocyte-conditioned medium. Interleukin-6 (IL-6) and interleukin-8 (IL-8) release from fibroblast cells was

Cellular in vitro Testing: Methods and Protocols
Edited by John W. Haycock, Arti Ahluwalia, and John M. Wilkinson

ISBN 978-981-4364-97-3 (Hardcover), 978-981-4364-98-0 (eBook)
www.panstanford.com

measured and used to confirm the early amplification stage of irritant-induced inflammatory signaling between keratinocytes and fibroblasts. We describe herein the methods to conduct these experiments in the context of paracrine models for in vitro testing of irritant chemicals at subtoxic concentrations, as an alternative to in vivo models.

7.1 Introduction

Since the 7th Amendment to the European Council (EC) Directive 76/768/EEC came into effect in 2009, the demand for new in vitro models for testing purposes has increased year on year. The EC directive imposes a marketing ban on products if chemicals therein have been tested on animals. In practice, this provides a strong direction for the use of alternative tests to be developed for irritation and specifically prohibits the use of animals for toxicological testing. Alternative models must therefore replace animals within Europe traditionally used for irritation, corrosivity, and phototoxicity tests. In parallel, the new European Registration, Evaluation and Authorisation of Chemicals (REACH) legislation requires all chemicals to be registered and labeled to establish an appropriate risk to human health. Approximately 30,000 tests will therefore need to be conducted to fulfill this requirement, traditionally with the Draize rabbit skin test being one of the most historically commonplace methods [1].

For skin, two reconstructed human models currently exist for in vitro irritation and corrosion testing. EpiDerm (produced by MatTek Ashland, MA) consists of normal human keratinocytes cultured on a collagen matrix. Similarly, EPISKIN (produced by SkinEthic & L'Oreal, Nice, France) consists primarily of normal human keratinocytes. Both models have been recently evaluated by the European Centre for the Validation of Alternative Methods (ECVAM), with EPISKIN being approved in April 2007 as an alternative to the rabbit Draize skin irritation test [2]. Such models have major improvements on previous two-dimensional (2D) models and are realistic alternatives to the Draize in vivo test.

The mechanism of skin irritation and the chemicals that trigger this response frequently cause an acute local inflammatory reaction

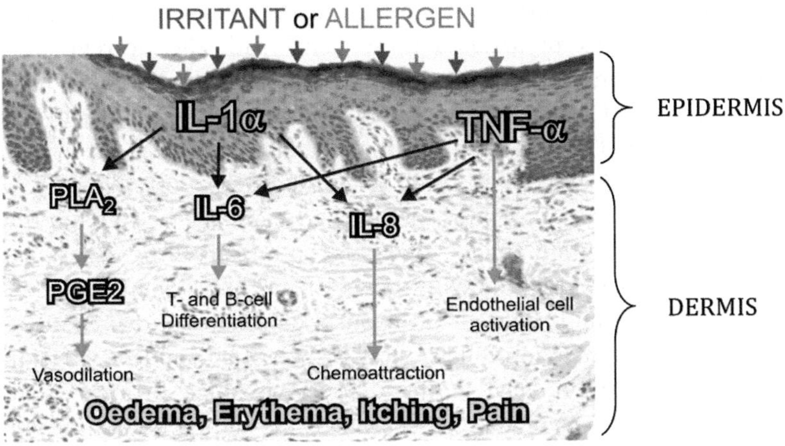

Figure 7.1 Epithelial-mesenchymal paracrine signaling in cutaneous irritation and inflammation in response to irritants or allergens. Adapted from Ref. 3, Copyright (2004), with permission from Elsevier.

without any direct or indirect cytotoxicity to the keratinocyte or fibroblast cells of the skin—reviewed in [3]. This particular mode of action highlights the need to develop in vitro models that are able to delineate between an inflammatory and a cytotoxicity response and is particularly important for compounds known to cause acute inflammation and contact dermatitis, which are not so easily detected by standard viability testing.

It is well described that if a chemical when acting as a skin irritant at subtoxic concentrations penetrates the stratum corneum [4], the viable and more basally located keratinocytes respond by releasing interleukin-1 alpha (IL-1α), initiating an acute inflammatory reaction [5–7] (Fig. 7.1). After a short period of time the IL-1α diffuses to the adjacent dermis, and an early amplification stage is triggered by the resident fibroblasts. Fibroblast cells express IL-1 receptors, and upon activation from keratinocyte-derived IL-1α, a secondary cytokine response comprised principally of IL-6, IL-8, tumor necrosis factor-alpha (TNF-α), and phospholipase A2 (PLA2) is then secreted by the fibroblasts [8]. The presence of these cytokines is required for continuing the inflammatory response, and each has a particular role. IL-8 acts as a chemoattractant for the local recruitment of T- and B-lymphocytes from the nearby vasculature, and IL-6 serves to activate these cells, with the up-regulation of adhesion molecule expression and activation of major histo compatibility (MHC) and inflammatory pathways. TNF-α activates the endothelial cells located within the local vasculature, facilitating the process of lymphocyte rolling and extravasation, while PLA2 increases the vascular permeability and capillary bed volume by causing vasodilation. Taken together, the tertiary process of local lymphocyte accumulation within the skin and subsequent pain, redness, swelling, and heat marks the onset of inflammation and an associated tissue reaction consistent with an irritant-induced reaction.

We have previously shown that a two-stage process exists whereby paracrine signaling occurs between human keratinocytes and fibroblasts grown in vitro in response to subtoxic concentrations of an irritant in the form of sodium dodecyl sulphate (SDS) [9]. Experimentally, we stimulated human keratinocytes as a first-stage inflammatory reaction with SDS and showed that they responded by secreting IL-1α. This was detected by the enzyme-linked immunosorbent assay (ELISA) and was also shown in the keratinocyte-conditioned medium to activate NF-kB as a second inflammatory stage in the fibroblast cells. The aim of the present work was therefore to confirm and extend this work, using the same two-stage paracrine-dependent process, asking whether irritant-induced keratinocytes activate dermal fibroblasts with an up-regulation of the secondary cytokines IL-6 and IL-8 in vitro, necessary for cytokine amplification and propagation of the inflammatory response.

7.2 Materials

7.2.1 *Human Keratinocyte and Fibroblast Isolation and Culture*

1. Human keratinocytes and human fibroblasts should be obtained under local ethical permissions and consent according to the country of study. Skin tissue is obtained typically from female patients undergoing abdominoplasty or breast reduction operations.

2. Green's medium consists of Dulbecco's modified Eagle's medium (DMEM) and Ham's F12 medium in a 3:1 ratio, supplemented with 10% fetal calf serum (FCS), 10 ng/mL epidermal growth factor (EGF), 0.4 μg/mL hydrocortisone, 0.18 mM adenine, 5 μg/mL insulin, 5 μg/mL transferrin, 2 mM glutamine, 0.2 μM triiodothyronine, 0.625 μg/mL amphotericin B, 100 IU/mL penicillin, and 100 μg/mL streptomycin (Sigma Chemical Company).
3. 0.1% (w/v) glucose (Sigma) is added.
4. Rat-tail collagen can be either purchased from Sigma or, alternatively, extracted directly under acidic conditions from rat tails.
5. Trypan blue (Sigma) is added.
6. Collagenase A (Boehringer Mannheim, Lewes, Sussex) is added.
7. DMEM plus 10% (v/v) neonatal calf serum (Advanced Protein Products), 2×10^{-3} mol/L glutamine, 0.625 μg/mL amphotericin B, 100 IU/mL penicillin, and 100 μg/mL streptomycin (Sigma) are added.
8. T75 and T25 flasks (Costar, High Wycombe, Bucks, U.K.) are used.
9. 0.1% w/v trypsin and 0.02% w/v ethylenediamine tetraacetic acid (EDTA) (Sigma) are added.

7.2.2 *Investigation of Paracrine Signaling between Keratinocytes and Fibroblasts with Activation of NF-kB*

1. 24-well plates (Costar, High Wycombe, Bucks, U.K.)
2. SDS (Sigma)
3. Potassium diformate (Formi®) (BASF, Germany)
4. Lipopolysaccharide (LPS) from *Escherichia coli* (serotype 0111:B4) (Sigma)
5. TNF-α (Roche, Welwyn Garden City, U.K.)

7.2.3 *Immunolabelling of NF-kB/p65*

1. Paraformaldehyde, phosphate buffered saline (PBS), and Triton-X100 (Sigma)
2. Anti-NF-kB/p65 (C-20) goat polyclonal immunoglobulin G (IgG) antibody (sc-109) (Santa Cruz Biotechnology, Santa Cruz, CA, USA)

3. Biotinylated anti-goat-IgG secondary antibody (BA-5000) (Vector Laboratories, CA, USA)
4. Fluorescein streptavidin (SA-5001) (Vector Laboratories)
5. Vectashield mounting medium containing propidium iodide (H-1300) (Vector Laboratories)
6. Epifluorescent or confocal microscope equipped with mean λ_{ex} = 494 nm / λ_{em} = 518 nm and λ_{ex} = 536 nm / λ_{em} = 617 nm

7.2.4 *Measurement of Interleukin-1 Alpha Production in Human Keratinocytes and Interleukin-6/8 Production by Fibroblasts*

1. ELISA kits for IL-1α, IL-6, and IL-8 (Quantikine; R&D Systems Europe Ltd., Abingdon, U.K.)

7.3 Methods

7.3.1 *Human Keratinocyte and Fibroblast Isolation and Culture*

1. The isolation of human epidermal keratinocytes and human dermal fibroblasts should be from skin obtained from abdominoplasty or breast reduction according to local ethically approved guidelines (in the present work, the NHS Trust, Sheffield, U.K.).
2. The method for keratinocyte isolation and culture was as previously described [10, 11]. Details are given below.
3. The method for the isolation and culture of fibroblasts was as previously described [12]. Details are given below.
4. Separate the dermis from the epidermis initially by removing subcutaneous tissue from the skin and cutting it into 0.5 cm^2 pieces. Incubate overnight at pH 7.45 at 4°C.
5. Peel the epidermal and dermal layers apart and collect the keratinocytes from the upper surface of the dermis and the underside of the epidermis using a scalpel. Pellet the cell suspension in Green's medium by centrifugation (1,000 g/5 min), wash in PBS by resuspension, and resuspend in Green's medium (20 mL).

6. Culture keratinocytes on rat-tail collagen–coated T75 flasks using Green's medium. Allow cells to attach to plastic wells coated with type I rat-tail collagen at 31 μg/cm^2 for approximately two days. It is preferable to use freshly isolated primary cultures for these studies, and when experiments are performed, a Trypan Blue cell dye exclusion test is useful to implement to confirm cellular viability.
7. Obtain fibroblasts from the dermis left after trypsinisation of split-thickness skin, as above. Washed the dermis in PBS and then mince finely with a scalpel and place it in 0.5% collagenase A overnight at 37°C.
8. Centrifuge the collagenase digest obtained from above (1,000 g/5 min) and discard the supernatant. Resuspend the fibroblast cell pellet in 4 mL of fibroblast culture medium: DMEM supplemented with 10% (v/v) neonatal calf serum, 2×10^{-3} mol/L glutamine, 0.625 μg/mL amphotericin B, 100 IU/mL penicillin, and 100 μg/mL streptomycin.
9. Place the samples in a T25 flask in an incubator (37°C/5% CO_2) until confluent. Passage the fibroblasts when 80–90% confluent into a T75 flask and use between passages 4 and 9.
10. When passaging fibroblasts use a 1 mL 1:1 mixture of 0.1% w/v trypsin and 0.02% w/v EDTA per T25 flask.

7.3.2 *Investigation of Paracrine Signaling between Keratinocytes and Fibroblasts with Activation of NF-kB*

1. Seed freshly isolated keratinocytes in 24-well plates at a density of 2×10^5 cells/well and culture for four to six days until approximately 60% confluent.
2. Separately seed fibroblasts in 24-well plates at a density of 2×10^5 cells/well and culture for four to six days until 80% confluent.
3. To investigate the direct effect of an irritant we have separately used two model chemicals, SDS and Formi®, for demonstrating the paracrine-dependent activation of NF-kB.
4. Cells should be incubated up to the highest nontoxic concentration with either SDS or Formi® and identified separately by viability measurement (e.g., by reduction tetrazolium dye 3-(4,5-

dimethylthiazol-2-yl)-2,5-diphenyltetrazolium bromide (MTT) to its insoluble formazan product). In our hands this is 10^{-4} M for SDS and 15 mM for Formi®. If other irritant chemicals are being determined, this should be conducted beforehand.

5. To investigate the paracrine activation of NF-kB in fibroblasts, incubate keratinocytes (at the above confluence) with SDS at 10^{-6}, 10^{-5}, or 10^{-4} M Formi® at 5, 10, and 15 mM for 2 h.
6. Collect the conditioned medium from the keratinocytes and add 1 mL to the 80% confluent monolayer of fibroblasts at a concentration of 100% for a further 2 h incubation period.
7. For the purposes of positive control, use LPS from *Escherichia coli* at 1 μg/mL or TNF-α at 1,000 IU/mL for a 2 h incubation period.
8. Negative controls samples should be incubated in Green's medium (keratinocytes) or DMEM (fibroblasts).

7.3.3 *Immunolabelling of NF-kB/p65*

1. Remove the medium from the fibroblasts and wash with PBS×3. Fix cells using 4% paraformaldehyde (in PBS) for 30 min, followed by permeablizing with Triton ×100 (0.1% w/v) for 20 min.
2. Wash thrice with PBS and block nonspecific protein sites with 5% dried-milk powder (in PBS) for 1 h.
3. Incubate with primary anti-NF-kB/p65 (C-20) goat polyclonal IgG antibody for 1 h at room temperature (1:100 v/v, in PBS).
4. Wash cells thrice with PBS and then incubate with biotinylated anti-goat-IgG secondary antibody (1:1000 v/v in PBS) for 1 h at room temperature.
5. Wash thrice with PBS and incubate samples with fluorescein streptavidin (1:100 v/v in PBS) for 30 min.
6. Either mount samples using coverslips (Vectashield mounting medium) containing propidium iodide to label nuclei (at 1.5 mg/mL) or observe directly after nuclei counterlabeling using an epifluorescence microscope with fluorescein-labeled NFkB/p65 ($\lambda_{ex} = 494$ nm / $\lambda_{em} = 518$ nm) and propidium iodide–labeled nuclei ($\lambda_{ex} = 536$ nm / $\lambda_{em} = 617$ nm) (Fig. 7.2).
7. All experiments should be performed in triplicate.

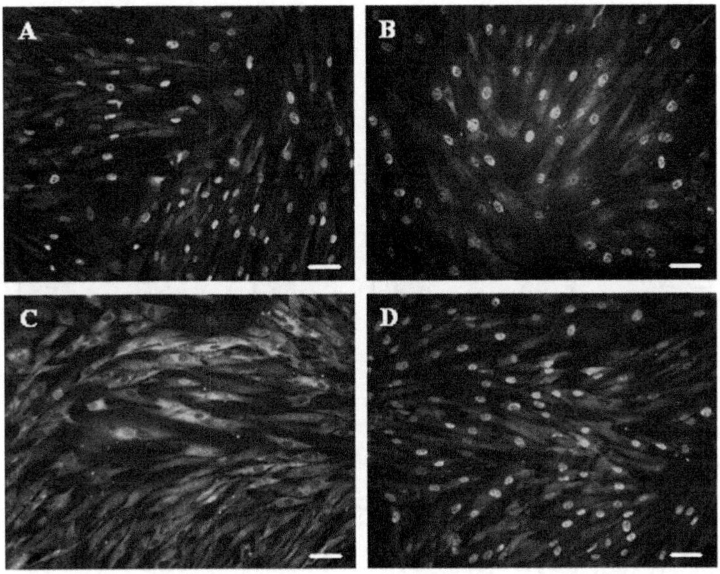

Figure 7.2 Immunolabeling of NF-kB/p65, showing paracrine activation in fibroblast cells by an SDS- or Formi®- incubated keratinocyte-conditioned medium. Fibroblast cells were incubated for 2 h with a conditioned medium from keratinocyte cells, which had previously been incubated for 2 h with (A) SDS (10^{-4} M) or (B) Formi® (15 mM). (C) Control nonincubated fibroblasts and (D) fibroblasts incubated with TNF-α (10 ng/mL, positive control). Scale bar = 50 μm. Reproduced with permission from [9].

7.3.4 *Measurement of Interleukin-1 Alpha Production in Human Keratinocytes and Interleukin-6/8 Production by Fibroblasts*

1. The detection of interleukins is best conducted using commercially available human-specific ELISA kits for IL-1α, IL-6, and IL-8.
2. Remove 100 μL of the keratinocyte-conditioned medium to measure IL-1α after a 2 h incubation with SDS or Formi® above according to the manufacturer's instructions (Fig. 7.3).
3. Remove 100 μL of the fibroblast-conditioned medium to measure IL-6 and IL-8 after a 2 h incubation with the keratinocyte-conditioned medium above according to the manufacturer's instructions (Figs. 7.4 and 7.5).

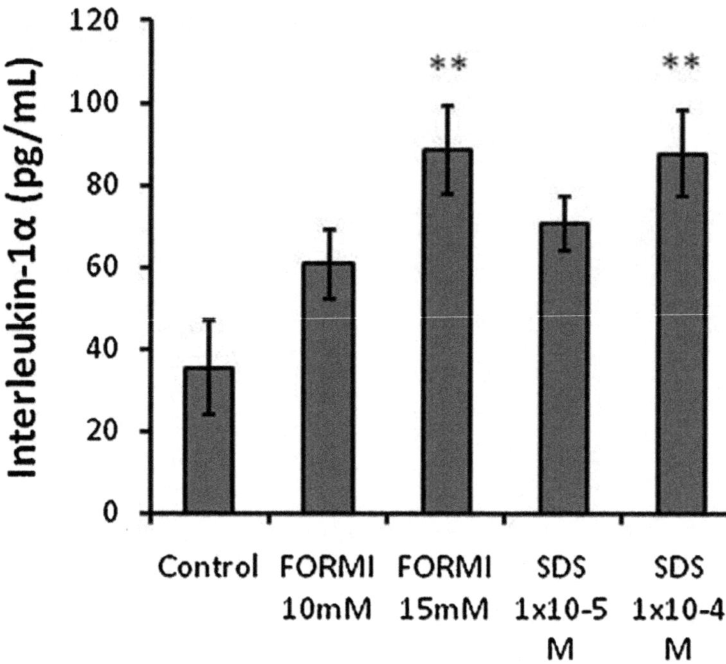

Figure 7.3 IL-1α production in a conditioned medium from keratinocyte cells incubated with SDS or Formi® by ELISA. Keratinocyte cells were stimulated for 2 h with Formi® (10 mM and 15 mM) or SDS (10^{-5} M and 10^{-4} M) vs. unstimulated control. Supernatants were collected in triplicate and IL-1α measured by a commercially available IL-1α ELISA. Values shown are means $\pm$ standard error of the mean ($n = 4$); $P < 0.05$, $P < 0.01$. Reproduced with permission from [9].

7.4 Notes

1. A dose response of the chemical to be tested using both keratinocytes and fibroblasts separately is recommended as a subtoxic concentration should be used. We previously found that IC50 of both SDS and Formi® for keratinocytes and fibroblasts by MTT viability measurement was similar [9].
2. If constitutively high levels of cytokines are detected in the absence of a potential irritant, we recommend removal of FCS from the culture medium during the incubation phase.

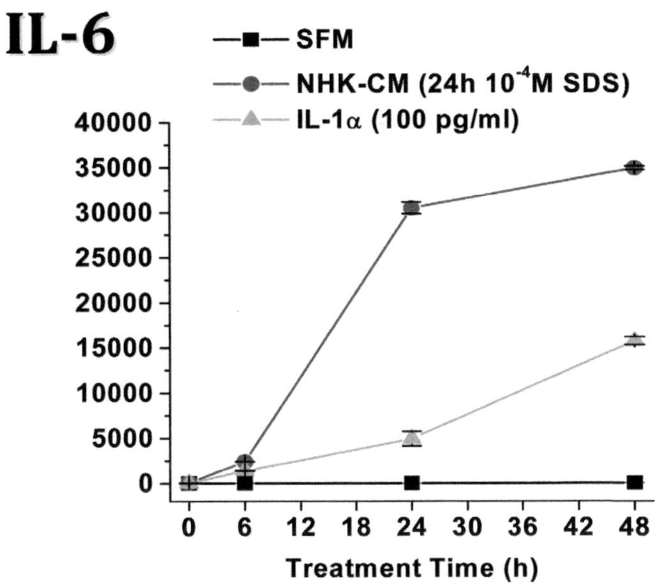

Figure 7.4 IL-6 production by fibroblast cells after incubation with a conditioned medium from keratinocyte cells previously incubated with SDS. Fibroblast cells were stimulated for 2 h with SDS (10^{-4} M) vs. unstimulated control. Supernatants were collected in triplicate and IL-6 measured by a commercially available ELISA. Values shown are means ± standard error of the mean ($n = 3$).

3. When analyzing NF-kB activation it is very important to introduce any potential activator when the cells are in the top part of log phase growth. This correlates with cell cultures being approximately 60% confluent.
4. NF-kB activation in a population of cells can be quantified, if required by introducing a simple scoring system whereby cytoplasmically located p65 = 0 (off), nuclear located p65 = 1 (on), and an equal distribution of p65 between the cytoplasm and nucleus = 0.5 (partially active). A minimum of 50 cell counts should be made for each condition and scores (numerator) divided between cells counted (denominator) to give a percentage activation of NF-kB per condition [9].

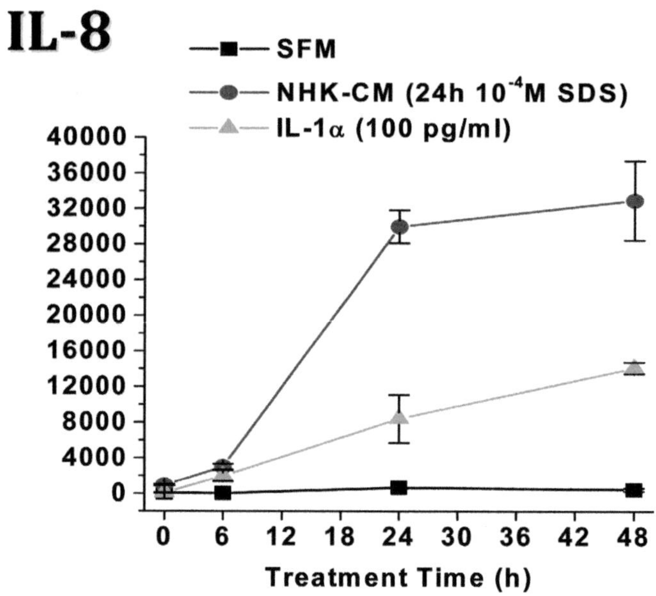

Figure 7.5 IL-8 production by fibroblast cells after incubation with a conditioned medium from keratinocyte cells previously incubated with SDS. Fibroblast cells were stimulated for 2 h with SDS (10^{-4} M) vs. unstimulated control. Supernatants were collected in triplicate and IL-8 measured by a commercially available ELISA. Values shown are means ± standard error of the mean ($n = 3$).

Acknowledgments

Jirapat Chunthapong was supported by a CHE-PhD-SFR scholarship from the Government of Thailand.

References

1. Draize, J. H., Woodard, G., Calvery, H. O. (1944) Methods for the study of irritation and toxicity of substances applied topically to the skin and mucous membranes. *J Pharmacol Exp Ther* **82**, 377–390.
2. de Brugerolle, A. (2007) SkinEthic laboratories, a company devoted to develop and produce in vitro alternative methods to animal use. *Altex-Altern Tierexp* **24**, 167–171.

3. Welss, T., Basketter, D. A., Schroder, K. R. (2004) In vitro skin irritation: facts and future. State of the art review of mechanisms and models. *Toxicol in Vitro* **18**, 231–243.
4. Ponec, M., Boxman, I., Aarden, L., Lowik, C. (1992) Keratinocyte-induced modulation of Il-6 production in fibroblasts. *J Invest Dermatol* **98**, 510–510.
5. Coquette, A., Berna, N., Vandenbosch, A., Rosdy, M., Poumay, Y. (1999) Differential expression and release of cytokines by an in vitro reconstructed human epidermis following exposure to skin irritant and sensitizing chemicals. *Toxicol in Vitro* **13**, 867–877.
6. Corsini, E., Galli, C. L. (1998) Cytokines and irritant contact dermatitis. *Toxicol Lett* **103**, 277–282.
7. Corsini, E., Bruccoleri, A., Marinovich, M., Galli, C. L. (1996) Endogenous interleukin-1 alpha is associated with skin irritation induced by tributyltin. *Toxicol Appl Pharm* **138**, 268–274.
8. Coquette, A., Berna, N., Vandenbosch, A., Rosdy, M., De Wever, B., Poumay, Y. (2003) Analysis of interleukin-1 alpha (IL-1 alpha) and interleukin-8 (IL-8) expression and release in in vitro reconstructed human epidermis for the prediction of in vivo skin irritation and/or sensitization. *Toxicol in Vitro* **17**, 311–321.
9. Canton, I., Cole, D. M., Kemp, E. H., Watson, P. F., Chunthapong, J., Ryan, A. J., MacNeil, S., Haycock, J. W. (2010) Development of a 3D human in vitro skin co-culture model for detecting irritants in real-time. *Biotech Bioeng* **106**, 794–803.
10. Goberdhan, N. J., Dawson, R. A., Freedlander, E., Macneil, S. (1993) A calmodulin-like protein as an extracellular mitogen for the keratinocyte. *Brit J Dermatol* **129**, 678–688.
11. Rheinwald, J. G., Green, H. (1975) Serial cultivation of strains of human epidermal keratinocytes—formation of keratinizing colonies from single cells. *Cell* **6**, 331–344.
12. Ralston, D. R., Layton, C., Dalley, A. J., Boyce, S. G., Freedlander, E., Mac Neil, S. (1999) The requirement for basement membrane antigens in the production of human epidermal/dermal composites in vitro. *Brit J Dermatol* **140**, 605–615.

Chapter 8

Advanced in vitro Models of the Intestinal Mucosa for Drug Delivery Studies

Eva-Maria Collnot,[a] Julia Susewind,[b] and Claus-Michael Lehr[a,b,c]

[a]*Department of Drug Delivery, Helmholtz Institute for Pharmaceutical Research Saarland, Saarland University, Saarbrücken, Germany*
[b]*Department of Biopharmaceutics and Pharmaceutical Technology, Saarland University, Saarbrücken, Germany*
[c]*Helmholtz Centre for Infection Research (HZI), Inhoffenstraße 7, 38124 Braunschweig, Germany*
lehr@mx.uni-saarland.de

Single–epithelial cell line models of biological barriers, such as the intestines or the lungs, have long been accepted as in vitro predictors of in vivo drug permeability. However, these models only reflect an isolated part of the epithelial barrier, neglecting other cell types integrated into the tissue in vivo, such as mucus-producing goblet cells or cells of the innate immune system. Furthermore, these cell lines represent the normal mucosa with intact tight junctions and strong barrier properties and do not take into account the patho-physiological changes happening in the case of inflammation, stress, or other tissue damage. Thus, they are not suitable to study targeting of drug delivery systems to diseased tissue or to investigate drug or drug carrier permeation across the susceptible barrier. Herein we

Cellular in vitro Testing: Methods and Protocols
Edited by John W. Haycock, Arti Ahluwalia, and John M. Wilkinson

ISBN 978-981-4364-97-3 (Hardcover), 978-981-4364-98-0 (eBook)
www.panstanford.com

describe a novel coculture setup of the intestinal mucosa, combining intestinal epithelial cells with a dendritic and a macrophage cell line, all three of human origin. The model can be advanced to a representative of the susceptible intestinal barrier by induction of inflammation with pro-inflammatory cytokine interleukin-1β. Forty-eight hours after stimulation, the model demonstrates major pathophysiological changes of the stressed intestinal barrier, such as decreased barrier function and release of biomarkers of inflammation. Therapeutic efficacy of drugs or formulations or possible harmful effects of xenobiotics can be monitored via measurement of transepithelial electrical resistance, cytokine measurement, and morphological investigations using confocal laser scanning microscopy.

8.1 Introduction

8.1.1 *Biological Barriers and Drug Permeation*

On the way from the application site to the biological target a drug is subject to four major steps of pharmacokinetics: **a**bsorption, **d**istribution, **m**etabolism, and **e**xcretion (the so-called ADME process). The interplay of these four processes will determine the extent to and the time after which a drug will appear at its site of action, thereby influencing the pharmacological effect. A drug's fate in the ADME process is interwoven with its ability to cross epithelia, a major biological barrier to drug distribution [1].

Epithelia line the cavities and surfaces of structures throughout the body and are characterized by the formation of confluent mono- or multilayers of polarized cells connected with each other by various cell–cell contacts such as gap junctions, adherent junctions, or tight junctions. In particular, the functionality of the latter type of cell contacts determines the tightness of the epithelial barrier, that is, the obstacle it represents to the permeation of a drug, xenobiotic or pathogen [2].

Specialized epithelial cells responsible for sensing or for secretion of fluids such as mucus may be found among the major population of simple epithelial cells. Furthermore, immune competent cells such as dendritic cells or tissue macrophages will often be associated with the epithelium located directly underneath the epithelial cells or invading across the epithelium into the lumen. They play a central role in tissue defense against invading microorganisms and particles and in the maintenance of immune homeostasis [3].

8.1.2 *Epithelial Cell Culture Models*

In healthy tissue, the absorption of a small molecule across the epithelium will mainly be limited by the tightness of the confluent monolayer of simple epithelial cells. Transport across an epithelial barrier is the sum of competing processes that influence the net absorption to varying extents. Depending on its physicochemical properties, mainly size and hydrophilicity, a molecule may transverse an epithelial barrier by two routes: (1) through the cells (transcellular transport) or (2) in between the cells (paracellular transport) via pores formed by the tight junctions. Transporter proteins, exhibiting low to high specificity, are integrated in the cell membrane and may modify transcellular transport, enabling otherwise poorly permeable drugs to cross the cell layer. By either influxing or effluxing their substrates transporters can increase or decrease overall compound permeability across an epithelial barrier [4, 5].

To study transport processes in vitro and to estimate drug permeability and thus bioavailability, cell culture models of simple epithelia are routinely used in ADME screening. A wide range of models exists for different tissues such as the intestinal epithelium, the bronchial or alveolar epithelium, the endothelium at the blood–brain barrier, or the cornea. Both primary cells and permanent cell lines (from cancer origin or immortalized by transfection) are used.

Grown on permeable cell culture inserts with polycarbonate or polyester membranes, epithelial cells form confluent cell mono- or multilayers, effectively separating the system into two compartments, that is, an apical compartment representing the lumen and a basolateral compartment representing the blood or tissue side. Transport of molecules or particles can be studied in this system in either direction, that is, absorptive transport in the apical to basolateral direction or secretory transport from the basolateral to the apical compartment. In this setup, the permeable membrane mimics the basement membrane and allows maximal differentiation of epithelial cells, thus giving the best approximate of the in vivo phenotype.

8.1.3 *Inflamed Coculture Model of the Intestinal Mucosa*

For the oral route, single–cell line models of enterocytes, such as the human colon carcinoma cell line Caco-2, are good approximates for the (small) intestinal mucosa with proven in vitro in vivo correlation for the transport of small molecules across the healthy tissue [6]. However, they are less suited to investigate absorption of large molecules or particles that often may be taken up by immune cells via phagocytosis [7]. Furthermore, the immune cells and their interaction with the epithelium play an essential role for inflammatory processes [8] and thus should be taken into account when looking at susceptible barrier models that mimic the pathophysiological changes occurring in the inflamed, stressed, or otherwise damaged tissue.

Covering this gap, we developed a triple coculture model of the intestinal mucosa combining Caco-2 enterocytes with THP-1 macrophages and MUTZ-3 dendritic cells. All cells are permanent cell lines and of human origin. Thus the model is an advancement on our previously established triple culture of Caco-2 cells and primary blood-derived macrophages and dendritic cells [9]. A reversible inflammation can be induced in the fully differentiated model by addition of pro-inflammatory cytokine interleukin-1β (IL-1β). The model then responds to the pro-inflammatory stimulus with a decrease but not total breakdown in epithelial barrier function and a release of soluble inflammation markers such as interleukin-8 (IL-8). Thus the model mimics major pathophysiological changes in the susceptible biological barrier in vivo.

8.2 Materials

1. Epithelial voltohmmeter (EVOM)
2. Human adenocarcinoma cell line Caco-2 C2Bbe1 (passage numbers 60 to 80)

3. Human macrophage–like cell line THP-1
4. Human dendritic–like cell line MUTZ-3
5. MUTZ-3 feeder cell line 5637
6. Well plates: 6
7. T75 cell culture flasks
8. #3640 Transwell plates with pore size 0.4 μm and 1.12 cm^2 diameter
9. Dulbecco's modified Eagle's medium (DMEM)
10. Roswell Park Memorial Institute (RPMI) 1640 medium
11. Alpha minimum essential medium (α-MEM)
12. Fetal calf serum (FCS)
13. Nonessential amino acids (NEAAs)
14. Penicillin/streptomycin (Pen/Strep)
15. Trypsin/ethylenediamine tetraacetic acid (EDTA)
16. Phosphate buffered saline (PBS) pH 7.4
17. Purecol collagen
18. Human AB serum
19. Phorbol 12-myristate 13-acetate (PMA)
20. IL-1β
21. CBA Human IL8 Flex set (BD Biosciences)

8.3 Methods

8.3.1 *Monocultures*

8.3.1.1 Caco-2

1. Cells are seeded in T75 cell culture flasks in DMEM supplemented with 10% FCS and 1% NEAA and incubated at 37°C and 5% CO_2.
2. The medium is changed every second day.
3. Cells are passaged every week. Thereto cells are washed twice with PBS. After that 3.5 mK Trypsin/EDTA is added to the cells, and cells are incubated for approximately 8 min at 37°C and 5% CO_2 until they are detached from the bottom of the flask. The reaction is stopped with a 6.5 mK cell culture medium. Cells are centrifuged for 4 min at 1,400 rpm. Afterward the pellet is resuspended in the cell culture medium. Cells are counted, and 0.5×10^6 cells are seeded in a new T75 cell culture flask.

8.3.1.2 THP-1

1. Cells are seeded in T75 cell culture flasks in the RPMI 1640 medium supplemented with 10% FCS and incubated at 37°C and 5% CO_2.
2. Cells are passaged every Tuesday and Friday. Thereafter cells are centrifuged for 4 min at 1,400 rpm. The cell pellet is resuspended in 10 mK of a fresh medium, and 3 mL of the cell suspension is transferred to a new T75 cell culture flask together with 10 mL of a fresh medium.
3. To stimulate the THP-1 cells to macrophages the cells are incubated with the THP-1 medium supplemented with PMA (5 ng/mL). After two days of incubation the cells adhere to the bottom of the cell culture flask and show macrophage-like morphology.

8.3.1.3 MUTZ-3

1. Cells are seeded at a density of 10^6 cells per mL in a six-well plate with α-MEM supplemented with 20% FCS and a 20% conditioned medium from 5,637 cells.
2. Twice a week 2 mL of the MUTZ-3 medium per well is exchanged for a fresh medium.
3. Then 5,637 cells are cultivated in the RPMI 1640 medium with 10% FCS. The cell supernatant is collected for three to four passages, filtered through a 0.22 μm cellulose filter, and stored as aliquots at –20°C for use in the MUTZ-3 cell culture medium.

8.3.2 *Setup of the Triple Culture*

1. 100 μL of dendritic cells (DCs) and macrophages are dispersed in a 80%(w/v) solution of type I bovine collagen, pH 7.4, at a concentration of 10^5 cells/mL and pipetted on top of 3460 Transwell filter inserts (1.12 cm^2 area), resulting in a seeding density of approximately 10^4 cells/well.
2. One hour later, 6×10^4 Caco-2 cells are seeded on top of the formed collagen gel layer (Fig. 8.1).
3. The coculture is kept at 37°C in 5% CO_2 with the Caco-2 cell culture medium with 1% Pen/Strep in the apical compartment

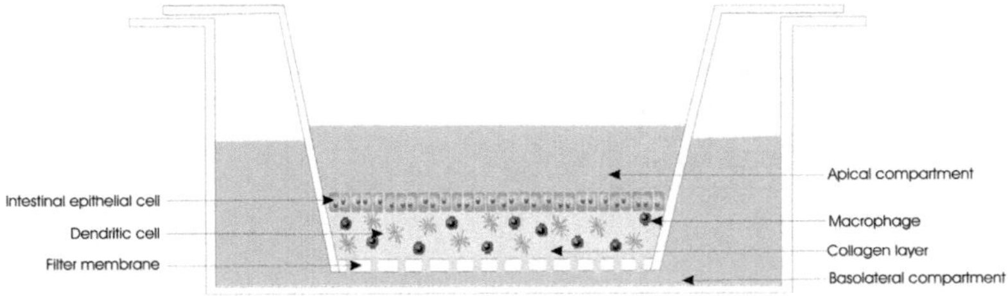

Figure 8.1 Setup of the triple coculture model of the intestinal epithelium. Inflammation is induced after 21 days in culture by addition of IL-1β to the apical compartment.

and the THP-1 medium with 1% Pen/Strep in the basolateral compartment.

4. After 21 days of coculture the Caco-2 cells form a confluent, fully differentiated epithelial monolayer, while the immune cells assume an intestinal phenotype [10]. Transepithelial electrical resistance (TEER) reaches values between 400 Ω/cm^2 and 600 Ω/cm^2 comparable to Caco-2 single cultures.

8.3.3 *Induction and Monitoring of Inflammation*

1. To induce a controlled inflammation, IL-1β (10 ng/mL) is added to the apical side of the coculture and incubated for at least two days.
2. Induction of full inflammation takes 24 h to 48 h. The process can be monitored by following release of pro-inflammatory markers IL-8 and tumor necrosis factor alpha (TNF-α) and by quantifying epithelial barrier breakdown via TEER measurement.
3. To determine IL-8 or TNF-α release, the cell culture medium (100–200 μL) is sampled from the apical and basolateral compartments of the coculture. Sampled volumes are replaced with a fresh medium.
4. IL-8 protein release into the supernatant is quantified by bead-based enzyme-linked immunosorbent assay (ELISA) (CBA Flex set) according to the manufacturer's protocol. Significantly increased IL-8 levels compared to nonstimulated cells or stimulated Caco-2 cells indicate successful induction of inflammation (Fig. 8.2).

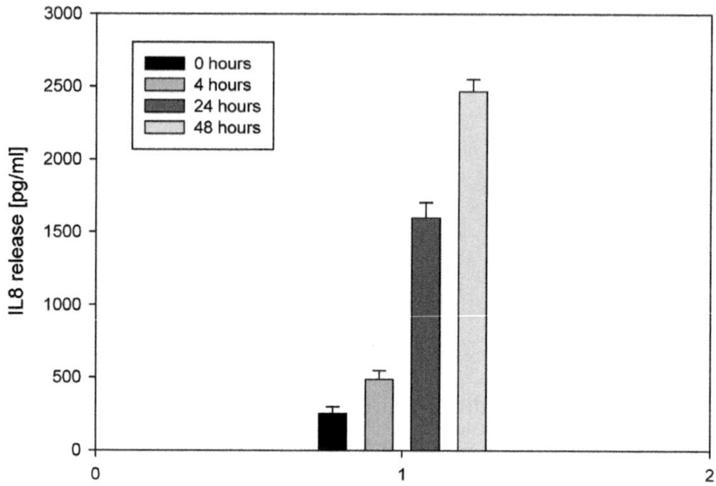

Figure 8.2 Cumulative release of IL-8 from the triple culture upon induction of inflammation by IL-1β.

5. To monitor rearrangement of tight junction proteins and subsequent reduction in barrier function, TEER is measured (see the chapter on nanotoxicity testing for further details). In the case of successful induction of inflammation, a time-dependent decline of TEER in the first 24 h can be observed, leveling off at a ~20% reduction in TEER compared to the nonstimulated triple culture after 48 h (Fig. 8.3).

8.3.4 *Testing in the Triple Culture Setup*

1. The apical fluid containing residual IL-1β is removed before the start of the experiment, and cells are washed three times with PBS.
2. Formulations/nanoparticles/substances to test are dissolved or dispersed in the Caco-2 cell culture medium +1% Pen/Strep and applied to the apical side of the coculture system.
3. Incubation time may depend on the test substance applied and the questions addressed (short time vs. repeated exposure).
4. Recovery or worsening of inflammation is monitored, as described above, via cytokine release and TEER measurement.

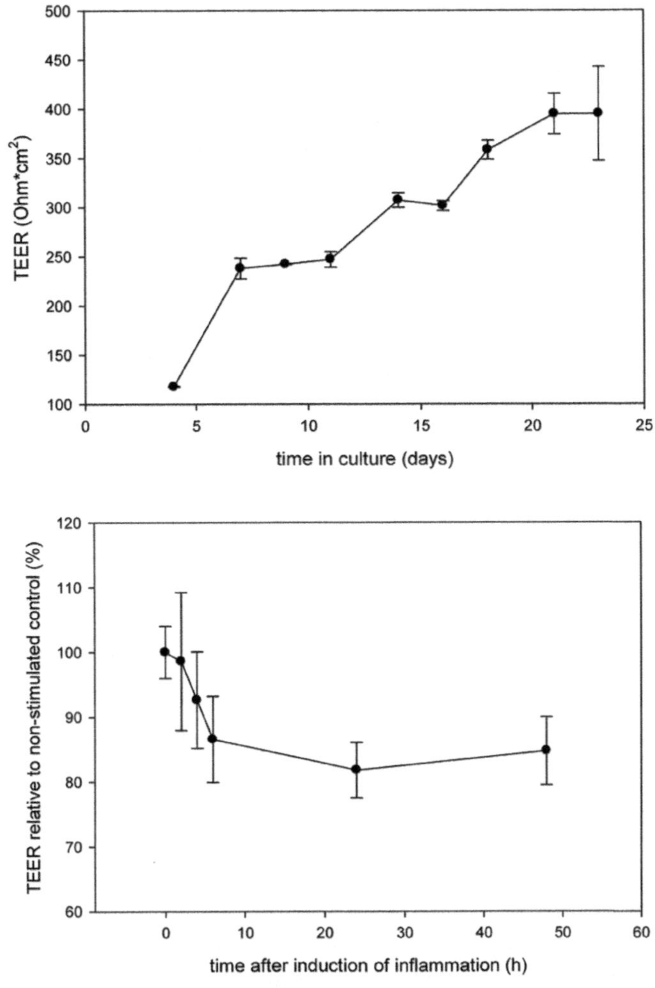

Figure 8.3 Development of TEER in the triple culture model during the first 21 days and upon induction of inflammation by IL-1β.

5. In nontreated control cells a self-healing effect will be observed as IL-8 release drops to noninflamed control levels approximately 72 h after removal of IL-1β. TEER recovery to healthy control levels takes 4 to 5 days in untreated control cells.
6. For fluorescent drugs or particles it is possible to study their disposition in the system and a possible preferential interac-

tion with either epithelial cells or immune cells via confocal microscopy. For this purpose Caco-2 cell nuclei are stained with 4′,6-diamidino-2-phenylindole (DAPI) (laser excitation: 720 nm multiphoton, emission: 390–465 nm), while immune competent cells show a strong red autofluorescence (laser excitation: 543 nm, emission: 560–615 nm).

8.4 Notes

1. For dispersing immune cells in the collagen solution to seed the triple culture, be sure to carefully adjust the pH of the solution to 7.4, as otherwise the vitality of the immune cells can be compromised. Therefore it's important to start by dispersing the immune cells in a cell culture medium with a serum, which has a certain buffering capacity, before adding the collagen solution (pH 2).
2. When pipetting the collagen solution with the immune cells into the wells it's important that there be no air bubbles in the collagen layer.
3. The Caco-2 cells must be seeded very carefully. Use an Eppendorf pipette and not a dispenser. Otherwise, the collagen layer is easy to destroy.
4. TEER is very sensitive to handling and temperature. Thus be quick but gentle in your handling of the cell culture plate and put the cell culture plate on a heating plate at 37°C during the measurement.
5. For exposure and follow-up studies lasting longer than 24 h, be sure to sterilize your formulation or test compound before applying it to the triple culture model. Even though the cell culture medium contains antibiotics this is not sufficient for long-term experiments if no other measures are taken.

Acknowledgments

This work was supported by the EC in the context of the FP7 project InLiveTox (CP-FP 228625-2).

References

1. Mälkiä, A., Murtomäki, L., Urtti, A., Kontturi, K. (2004) Drug permeation in biomembranes: in vitro and in silico prediction and influence of physicochemical properties. *Eur J Pharm Sci* **23**, 13–47.
2. Marchiando, A. M., Graham, W. V., Turner, J. R. (2010) Epithelial barriers in homeostasis and disease. *Annu Rev Pathol—Mech Dis* **5**, 119–144.
3. MacDonald, T. T., Monteleone, I., Fantini, M. C., Monteleone, G. (2011) Regulation of homeostasis and inflammation in the intestine. *Gastroenterology* **140**, 1768–1775.
4. Leslie, E. M., Deeley, R. G., Cole, S. P. (2005) Multidrug resistance proteins: role of P-glycoprotein, MRP1, MRP2, and BCRP (ABCG2) in tissue defense. *Toxicol Appl Pharmacol* **204**, 216–237.
5. DuBuske, L. M. (2005) The role of P-glycoprotein and organic anion-transporting polypeptides in drug interactions. *Drug Safety* **28**, 789–801.
6. Sun, H., Chow, E. C., Liu, S., Du, Y., Pang, K. S. (2008) The Caco-2 cell monolayer: usefulness and limitations. *Expert Opin Drug Metab Toxicol* **4**, 395–411.
7. Lavelle, E. C., Sharif, S., Thomas, N. W., Holland, J., Davis, S. S. (1995) The importance of gastrointestinal uptake of particles in the design of oral delivery systems. *Adv Drug Delivery Rev* **18**, 5–22.
8. Maldonado-Contreras, A. L., McCormick, B. A. (2011) Intestinal epithelial cells and their role in innate mucosal immunity. *Cell Tissue Res* **343**, 5–12.
9. Leonard, F., Collnot, E. M., Lehr, C. M. (2010) A three-dimensional coculture of enterocytes, monocytes and dendritic cells to model inflamed intestinal mucosa in vitro. *Mol Pharm* **7**, 2103–2119.
10. Wang, F., et al. (2006) IFN-γ-induced TNFR2 expression is required for TNF-dependent intestinal epithelial barrier dysfunction. *Gastroenterology* **131**, 1153–1163.

Chapter 9

An in vitro NHBE Model of the Human Bronchial Epithelium

Zoë Prytherch and Kelly BéruBé
School of Biosciences, Cardiff University, Museum Avenue, Wales, CF10 3AX, U.K.
prytherchzc@cf.ac.uk

The respiratory tract is the primary site of exposure to airborne substances, with the bronchial epithelium providing the initial line of defense. In addition, the disruption and/or remodeling of the bronchial epithelium occurs as a result of numerous airway diseases such as asthma and chronic obstructive pulmonary disorder (COPD). With the burden of lung disease ever increasing, the principle focus in respiratory medicine is to gain a better understanding of the mechanisms of lung injury/repair as well as the development of more effective therapeutics. Traditionally animal models have been the primary focus in respiratory research; however, they are deficient in many key areas in their ability to mimic the human response to inhaled compounds. A growing need therefore exists for an accurate, in vitro model of the bronchial epithelium, which accurately reflects the human in vivo situation. We have created a normal human bronchial epithelial (NHBE) model of the bronchial epithelium cultured at the air–liquid interface (ALI) in Millipore®

Cellular in vitro Testing: Methods and Protocols
Edited by John W. Haycock, Arti Ahluwalia, and John M. Wilkinson

ISBN 978-981-4364-97-3 (Hardcover), 978-981-4364-98-0 (eBook)
www.panstanford.com

cell culture inserts. This NHBE model is fully differentiated, pseudostratified, mucociliary epithelium containing basal, serous, Clara, goblet, and ciliated cells. The NHBE model is stable and fully developed from days 24–33 in an ALI culture.

9.1 Introduction

Inhalation exposes the lung to a variety of substances [1, 2], with the bronchial epithelium providing one of the initial lines of defense to such inhaled particles/gases [3, 4]. In addition to this a variety of airway diseases (e.g., chronic obstructive pulmonary disorder [COPD] and asthma) result in damage and subsequent remodeling of the respiratory epithelium [5–7]. Therefore the bronchial epithelium is an ideal region to study both the toxicological consequences to inhaled substances and the pathology of and/or candidate respiratory drugs for airway diseases.

Historically, animals, especially mammals, with pulmonary anatomy most similar to that of humans are used in respiratory research [8–10]. Animal models possess both similarities and differences to the human in vivo situation. Arguments in favor of in vivo, instead of in vitro, models include the fact that they produce a systemic effect to the response against foreign debris. However, there are numerous discrepancies between humans and animals that affect the accuracy of animal models, including particle deposition [11], anatomical differences in airway branching patterns [12], types and composition of cells present [13, 14], differences in biotransforming enzymes [15], and physiological differences such as changes in breathing patterns and reduction in metabolic rates [18]. Therefore, a degree of adaptation and extrapolation must be undertaken to obtain the actual human in vivo response, which is far from ideal. On the basis of this, many researchers are striving for increasingly accurate models to predict human pulmonary toxicity. In addition, there is a need for the development of alternative, yet comparable, in vitro models to refine, replace, and reduce the number of animals used in scientific research [17–19]. For these and other reasons we decided to pursue an in vitro model from human primary cells.

Human cell culture models are currently the main focus of in vitro respiratory models and can be divided into three main groups:

(1) Carcinoma-derived, such as the Calu-3s [20]. These form confluent, polarized monolayers that secrete mucus glycoproteins and contain microvilli [21], possess transepithelial electrical resistance (TEER; with discrepancies between laboratories) [22], and contain P450 enzymes [23], transport systems, and efflux pumps [24]. Limitations of the Calu-3 model includes a lack of pseudostratification and cilia and deficiencies in various biotransforming enzymes, along with some uncharacteristic phenotypes owing to the carcinogenic origin of cells [20, 21].
(2) Virus-transformed cell lines, such as the 16HBE14o- cells: human bronchial epithelial cells transformed by SV40 [25]. These were initially cultured to examine the cystic fibrosis conductance regulator (CFTR) [26]. The 16HBE14o- cultures form polarized, multilayered cultures, possessing tight junctions and numerous drug transporters and drug-related proteins [20–28]. The main downside to this cell line include a deficiency in producing a protective mucus covering and alteration or nonexpression of important transcription factors for important genes [20].
(3) Primary cells, such as the normal human bronchial epithelial (NHBE) cells (used here). These are nontransformed cells that form a well-differentiated epithelium of mucociliary phenotype and reflect the human in vivo bronchial epithelium. However, due to the primary nature of NHBE cells the major drawback of this model is its limited culture life span. Despite this, we believe that the NHBE model is a superior in vitro model of the bronchial epithelium [29].

To accurately recreate a human-based in vitro model of the bronchial epithelium certain parameters need to be met. First, the use of human primary cells (NHBEs) was deemed essential to create an in vivo–like model reflective of the human situation. Second was the selection of a biomimetic scaffold for our filter-well bioreactors or "inserts" to allow culturing at the air–liquid interface (ALI), which is essential to allow polarization and differentiation of the epithelium [30]. Third is the presence of specific growth factors and hormones, specifically retinoic acid, in the culture medium [31]. Here we will demonstrate how to culture an in vitro human bronchial epithelium model that morphologically resembles the human in vivo epithelium and contains basal cells (BCs), serous cells, Clara cells (ClCs), goblet cells (GCs), and ciliated cells (CCs) [29].

9.2 Materials

9.2.1 *Initial Seeding and Culture to Confluence of NHBE Cells*

1. NHBE cells are taken, cryopreserved in Bronchial Epithelial Growth Medium® (BEGM®), with retinoic acid (Lonza Group, Switzerland).
2. BEGM® was prepared with commercially purchased (Lonza Group, Switzerland) bronchial epithelium basal medium (BEBM) and BEGM® SingleQuots® (10 mL BEGM: 10 mL BEBM, 40 μL bovine pituitary extract, and 10 μL of insulin, hydrocortisone, transferrin, triodothyronine, epinephrine, epidermal growth factor, gentamicin/amphoteracin, and retinoic acid).
3. T75 cell culture flasks (Fisher Scientific, Manchester, U.K.) are taken.

9.2.2 *Second Passage*

1. Hank's balanced salt solution (HBSS) (Invitrogen, Paisley, Scotland)
2. Trypsin–ethylenediamine tetraacetic acid (EDTA) (Invitrogen, Paisley, Scotland)
3. 3 mL foetal calf serum (FCS) (JRH BioScience, Kansas, USA) added to 27 mL RPMI-1640 (Invitrogen, Paisley, Scotland) to create a 10% FCS neutralization solution
4. T175 cell culture flasks (Fisher Scientific, Manchester, U.K.)

9.2.3 *Seeding into Cell Culture Inserts*

1. Millicell® hanging cell culture insert, polyethylene terephthalate: 6.5 mm diameter, 0.4 μm pore size (Millipore®, Watford, U.K.)

was used (see Note 1) for all but 6x cell culture inserts. The other 6x inserts were Costar® Transwell® cell culture inserts; 6.5 mm, 0.4 μm pore size (Corning Life Sciences, Schipol-Rijk, the Netherlands) as the Transwell® inserts allow visualization of the cells using phase-contrast microscopy.
2. A 24-well cell culture plate (Grenier Bio-one, U.K.) was used.

9.2.4 *Confluent Insert Culture Inserts*

1. To make 20 mL ALI media: 10 mL BEBM, 10 mL Dulbecco's modified Eagle's medium (DMEM), 80 μL bovine pituitary extract, and 20 μL of insulin, hydrocortisone, transferrin, triodothyronine, epinephrine, epidermal growth factor, gentamicin/amphoteracin, and retinoic acid

9.2.5 *Transepithelial Electrical Resistance*

1. ENDOHM-6 Endohm Chamber (World Percision Instruments, Stevenage, U.K.)
2. Phosphate buffered saline (PBS), pH 7.4, with $CaCl_2$ or $MgCl_2$ (Invitrogen, Paisley, Scotland)
3. An epithelial voltohmmeter (EVOM) (World Percision Instruments, Stevenage, U.K.)

9.3 Methods

9.3.1 *Initial Seeding and Culture to Confluence of NHBE Cells*

1. A cryovial of NHBE cells (~500,000 cells) was rapidly (2–3 min) thawed (37°C) and suspended in BEGM® (37°C). The NHBE–BEGM® mixture (15 mL) was transferred to T75 flasks and incubated overnight.
2. The following day (see Note 2) the BEGM® (15 mL) was replaced with a 15 mL fresh BEGM® medium (37°C), and this was then replaced (20 mL) every other day until the cells reached 70% confluence (~2–3 days).

9.3.2 *Second Passage*

1. Confluent cell cultures were rinsed twice with 5 mL HBSS (37°C).
2. Trypsinization of cells was achieved by addition of 3 mL Trypsin-EDTA solution for three to 5 min (37°C) until 90% of the cells were detached.
3. FCS (10%) in RPMI (30 mL) was added to neutralize the trypsin, and the cell suspension was removed and centrifuged (72 × g for 4 min). The supernatant was removed, and the cell pellet was resuspended in a total of 2 mL BEGM® (37°C).
4. The resuspended cell pellet (2 mL) was added to the remaining BEGM® (37°C), and the BEGM®-cell suspension (30 mL) was placed in each T175 cell culture flask and incubated overnight.
5. The subsequent day (see Note 2) the BEGM® (30 mL) was replaced with a 30 mL fresh BEGM® medium (37°C). BEGM® (30 mL) was replaced every third day until the cells reached 70% confluence (~2–3 days).

9.3.3 *Seeding into Cell Culture Inserts*

1. Steps 1 and 2 from Section 9.3.2 are repeated.
2. FCS (10%) in RPMI (30 mL) was added to neutralize the trypsin, and the cell suspension was removed and centrifuged (72 × g for 4 min). The supernatant was aspirated, and the cell pellets were resuspended and combined in a total of 2 mL BEGM® (37°C).
3. The cell concentration of the BEGM®-cell suspension was determined by using a 5 × 5 grid on a hemocytometer. The required seeding density for imputing into inserts was 0.5 million cells/mL.
4. The BEGM®-cell suspension was added to the required volume of BEGM® for seeding, and 150 μL of the BEGM®-cell suspension was placed apically into each cell culture insert, with 500 μL of BEGM® placed in each well (beneath the insert) of the 24-well cell culture plate. Cultures were incubated overnight (see Note 3).
5. NHBE cells were seeded into Millipore® inserts, as well as into 6x Transwell® inserts, as Transwell® inserts allow visualization of the cells using phase-contrast microscopy (Fig. 9.1).

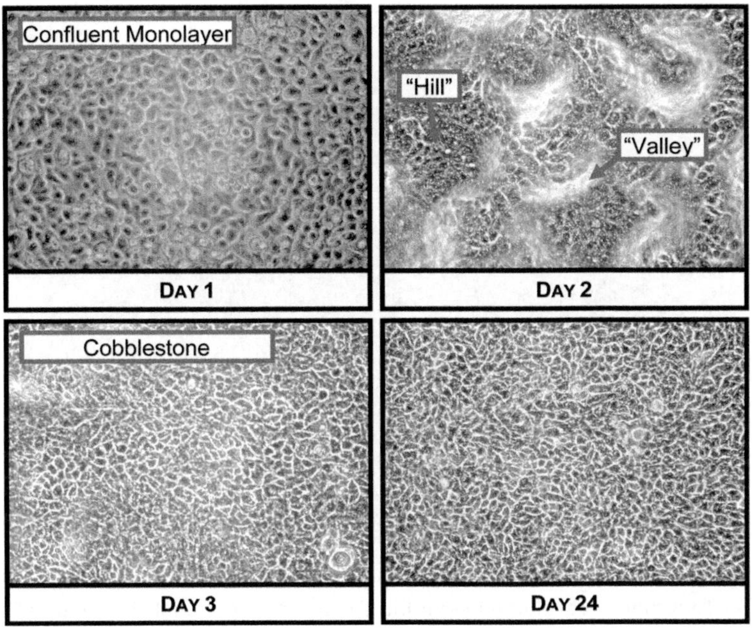

Figure 9.1 Phase-contrast light microscopy images of NHBE cells grown in Transwell® inserts. Days are postseeding into inserts.

9.3.4 *Confluent Insert Cultures*

1. 24 h after seeding into cell culture inserts the apical medium was removed to create an ALI culture (day 1). The basal medium (BEGM®) was replaced with 300 μL of the ALI medium. The basal ALI medium was changed daily, adding 300 μL of a fresh ALI medium into each well for the first week and then six days a week thereafter.
2. Once the cells began to show mucin secretion (see Note 4), it was carefully removed by aspiration every third day to prevent mucin accumulation (as there was no mucociliary escalator to remove the mucin).
3. The NHBE cultures were cultured in this way until fully differentiated and stable (day 24) and could be utilized for morphological analysis or toxicological testing up to and including day 33 (Fig. 9.2) (see Note 5).

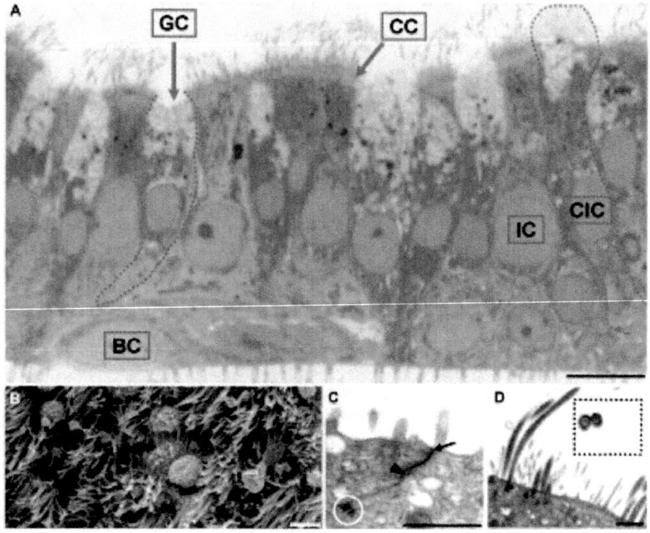

Figure 9.2 Images of the fully differentiated NHBE model (days 24–33). (A) Toluidine blue–stained semithin section of the model. The epithelium is pseudostratified and contains BCs, ICs, GCs, CCs, and ClCs. Scale bar = 10 μm. (B) SEM image displaying the surface topography of the NHBE culture. Scale bar = 5 μm. (C) TEM denoting the presence of tight junctions (arrow), adherens junctions (arrowhead), and desmosomes (circled). Scale bar = 1 μm. (D) TEM displaying the presence of cilia and anchoring basal bodies. (Inset) Cross section of cilia displaying 9 + 2 axoneme formation. Scale bar = 1 μm. *Abbreviation*: IC, intermediate cell.

9.3.5 *Transepithelial Electrical Resistance*

1. TEER is an important biochemical parameter to assess junctional formation in cell cultures and is considered a valid and sensitive method [32, 33]. The TEER values of cell cultures were evaluated over days 1–42 in the ALI culture using the Endohm chamber.
2. The chamber was filled with 1 mL of PBS solution, immersing the bottom electrode. A cell insert was then placed inside the Endohm chamber, and the PBS solution (150 μl) was placed on the apical region of the cultures (i.e., into the insert). The top electrode (lid) was placed on top of the insert, and the TEER reading was taken by pressing the "current release" button on the EVOM, releasing the current.

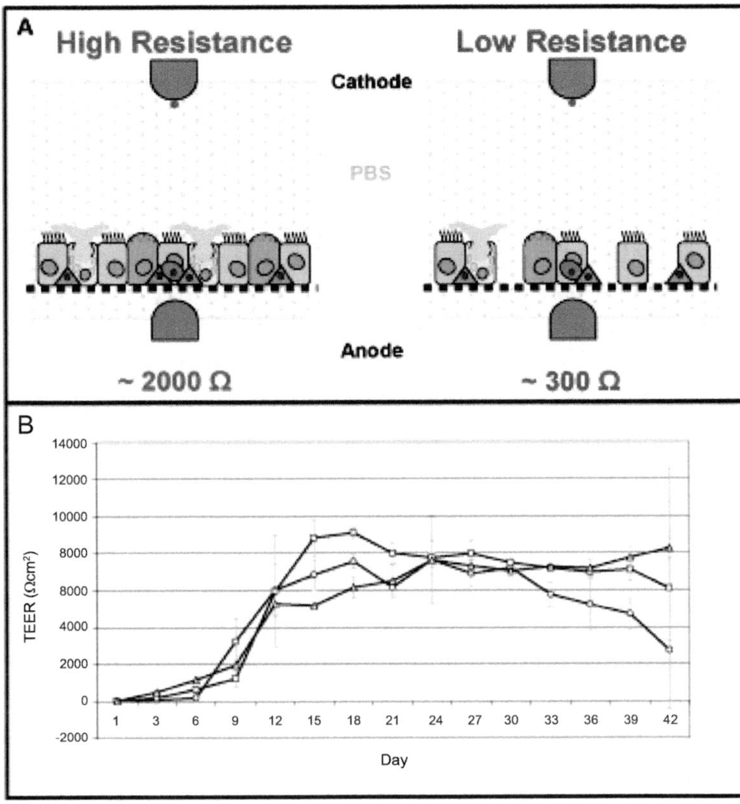

Figure 9.3 Image to explain the TEER of NHBE cultures. (A) Schematic diagram of NHBE cultures inside the Endohm chamber. The diagram on the left depicts a "healthy," full-differentiated culture with functional junctions, resulting in a high TEER reading. The diagram on the right denotes a disrupted epithelium, with a associated low TEER reading. (B) Graph displaying the TEER values of three different NHBE donors cultured at the ALI for 42 days. □: Donor A; ○: Donor B; Δ: Donor C. Error bars denote standard deviation; $n = 3$.

3. The EVOM detected the strength of the current reaching the bottom electrode, converting it into the resistance (Ω) created by the insert (Fig. 9.3). Prior to measuring the resistance values for the cell cultures, a "blank" reading was recorded by using an insert without cells (with PBS as above) and was later subtracted from that of the cell culture reading.

9.4 Notes

1. During our initial optimization of the model we tested numerous cell culture inserts, and we found that the Millicell® hanging cell culture insert with the polyethylene terephthalate membrane, a diameter of 6.5 mm and a 0.4 μm pore size, produced the best morphological and biochemical results. The only disadvantage to these inserts we found were that the membrane was not transparent and therefore the cultures could not be viewed by phase-contrast microscopy. To overcome this we also seeded 6x Costar® Transwell® cell culture inserts, 6.5 mm diameter and 0.4 μm pore size. These inserts were only used as a visual check of culture viability. We would advise either using the same inserts we use or testing out other inserts that you may want to use. If you do this please use microscopy and TEER to confirm that your cultures are morphologically and biochemically healthy.
2. A phase-contrast microscope should be used after overnight incubation to check the attachment and shape (elongated oval) of cells after seeding in flasks or cell culture inserts (Transwell® inserts only; see Fig. 9.1).
3. As you do not want to leave the cell suspension for too long in the universal container prior to seeding, it is important to do as much preparation as possible prior to trypsinizing the cells. This will include placing the cell culture inserts into the 24-well plates and placing 500 μL of BEGM® (37°C) in each well below the insert.
4. Cultures typically begin to produce mucin around day 15–18 after seeding into inserts. The mucin can be observed as a layer of clear viscous liquid on the apical surface of the cultures.
   N.B. If the apical liquid is nonviscous and pink in color then this is not mucin but the medium leaking through the culture. This indicates that the culture is no longer healthy, and if a TEER measurement was taken it would be very low. This observation only holds true after day 24 in inserts.
5. As a routine check we measure the TEER of cultures immediately prior to utilizing in any experiments in order to ensure that the cultures are healthy.

Acknowledgments

The authors would like to thank AstraZeneca R&D, U.K., for funding this work.

References

1. Cloutier, M. M., Thrall, R. S. (2004) The respiratory system. In Berne, R. M., Koeppen, B. M., Stanton, B. A. (eds.), *Physiology*, 5th ed., St. Louis, MO: Mosby, 443–535.
2. Korpáš, J., Honda, Y. (1996) Aspects of airway defence mechanisms. *Pathophysiology* **3**, 81–86.
3. Gerritsen, J. (2000) Host defence mechanisms of the respiratory system. *Paediatr Respir Rev* **1**, 128–134.
4. Nicod, L. P. (1999) Pulmonary defence mechanisms. *Respiration* **66**, 2–11.
5. Puchelle, E., Zahm, J. M., Tournier, J. M., Coraux, C. (2006) Airway epithelial repair, regeneration, and remodeling after injury in chronic obstructive airway disease. *Proc Am Thorac Soc* **3**, 726–733.
6. Trautman, A., Krüger, K., Akdis, M., Müller-Wening, D., Akkaya, A., Bröcker, E.-B., Blasser, K., Akdis, C. A. (2005) Apoptosis and loss of adhesion of bronchial epithelial cells in asthma. *Int Arch Allergy Immunol* **138**, 142–150.
7. White, S. R., Dorscheid, D. R. (2002) Corticosteroid-induced apoptosis of airway epithelium: a potential mechanism for chronic airway epithelial damage in asthma. *Chest* **122**, 278–284.
8. Chu, C. H., Liu, D. D., Hsu, Y. H., Lee, K. C., Chen, H. I. (2007) Propofol exerts protective effects on the acute lung injury induced by endotoxin in rats. *Pulm Pharmacol Ther* **20**, 503–512.
9. Fiala, E. S., Sohn, O. S., Wang, C. X., Seibert, E., Tsurutani, J., Dennis, P. A., El-Bayoumy, K., Sodum, R. S., Desai, D., Reinhardt, J., Aliaga, C. (2005) Induction of preneoplastic lung lesions in guinea pigs by cigarette smoke inhalation and their exacerbation by high dietary levels of vitamins C and E. *Carcinogenesis* **26**(3), 605–612.
10. Lin, S., Walker, J., Xu, L., Gozal, D., Yu, J. (2007) Behaviours of pulmonary sensory receptors during development of acute lung injury in the rabbit. *Exp Physiol* **92**, 749–755.

11. Schlesinger, R. B. (1985) Comparative deposition of inhaled aerosols in experimental animals and humans: a review. *J Toxicol Environ Health* **15**, 197–214.
12. Brody, A. R. (1984) Inhaled particles in human disease and animal models: use of electron beam instrumentation. *Environ Health Perspec* **56**, 149–162.
13. Breeze, R. G., Wheeldon, E. B. (1977) The cells of the pulmonary airways. *Am Rev Respir Dis* **116**, 705–777.
14. Thomassen, D. G., Nettesheim, P. (1990) *Biology, Toxicology, and Carcinogenesis of Respiratory Epithelium*, London, UK: Hemisphere Publishing Corporation.
15. Castell, J. V., Donato, M. T., Gomez-Lechon, M. J. (2005) Metabolism and bioactivation of toxicants in the lung. The in vitro cellular approach. *Exp Toxicol Pathol* **57**, 189–204.
16. Mautz, W. J. (2003) Exercising animal models in inhalation toxicology: interactions with ozone and formaldehyde. *Envir Res* **92**, 14–26.
17. EU Directive 76/768/EEC. (2003) Directive 2003/15/EC of the European Parliament and of the Council of 27 February 2003. *Official J Eur Union* **L66**, 26–35.
18. http://www.frame.org.uk. Date last accessed: Dec 14, 2010.
19. http://www.nc3rs.org.uk. Date last accessed: Dec 14, 2010.
20. Forbes, B. (2000) Human airway epithelial cell lines for in vitro drug transport and metabolism studies. *Pharm Sci Technol Today* **3**(1), 18–27.
21. Grainger, C. I., Greenwell, L. L., Lockley, D. J., Martin, G. P., Forbes, B. (2006) Culture of Calu-3 cells at the air interface provides a representative model of the airway epithelial barrier. *Pharm Res* **23**(7), 1482–1490.
22. BéruBé, K., Aufderheide, M., Breheny, D., Clothier, R., Combes, R., Duffin, R., Forbes, B., Gaça, M., Gray, A., Hall, I., Kelly, M., Lethem, M., Liebsch, M., Merolla, L., Morin, J. P., Seagrave, J., Swartz, M. A., Tetley, T. D., Umachandran, M. (2009) In vitro models of inhalation toxicity and disease: the report of a frame workshop. *ATLA, Altern Lab Anim* **37**(1), 89–141.
23. Foster, K. A., Oster, C. G., Mayer, M. M., Avery, M. L., Audus, K. L. (2000) Characterization of the Calu-3 cell line as a tool to screen pulmonary drug delivery. *Int J Pharm* **208**, 1–11.
24. Florea, B. I., Cassara, M. L., Junginger, H. E., Borchard, G. (2003) Drug transport and metabolism characteristics of the human airway epithelial cell line Calu-3. *J Control Rel* **87**, 131–138.

25. Westmoreland, C., Walker, T., Matthews, J., Murdock, J. (1999) Preliminary investigations into the use of a human bronchial cell line (16HBE14o-) to screen for respiratory toxins in vitro. *Toxicol in Vitro* **13**, 761–764.

26. Gruenert, D. C., Basbaum, C. B., Welsh, M. J., Li, M., Finkbeiner, W. E., Nadel, J. A. (1988) Characterization of human tracheal epithelial cells transformed by an origin-defective simian virus 40. *Proc Natl Acad Sci USA* **85**, 5951–5955.

27. Ehrhardt, C., Kneuer, C., Laue, M., Schaefer, U. F., Kim K-J., Lehr, C.-M. (2003) 16HBE14o- human bronchial epithelial cell layers express p-glycoprotein, lung resistance-related protein, and caveolin-1. *Pharm Res* **20**(4), 545–551.

28. Wan, H., Winton, H. L., Soeller, C., Stewart, G. A., Thompson, P. J., Gruenert, D. C., Cannell, M. B., Garrod, D. R., Robinson, C. (2000) Tight junction properties of the immortalized human bronchial epithelial cell lines Calu-3 and 16HBE14o. *Eur Respir J* **15**, 1058–1068.

29. Prytherch, Z., Job, C., Marshall, H., Oreffo, V., Foster, M., BéruBé, K. (2011) Tissue-specific stem cell differentiation in an in vitro airway model. *Macromol Biosci* **11**(11), 1467–1477.

30. Wu, R., Sato, G. H., Whitcutt, M. J. (1986) Developing differentiated epithelial cell cultures: airway epithelial cells. *Fund Appl Toxicol* **6**, 580–590.

31. Gardner, T. W., Lieth, E., Khin, S. A., Barber, A. J., Bonsall, D. J., Lesher, T., Rice, K., Brennan, W. A. Jr. (1997) Astrocytes increase barrier properties and zo-1 expression in retinal vascular endothelial cells. *Invest Opthalmol Vis Sci* **38**(11), 2423–2427.

32. Pasternak, A. S., Miller, W. M. (1995) Measurement of trans-epitheliai electrical resistance in perfusion: potential application for in vitro ocular toxicity testing. *Biotechnol Bioeng*, **50**(5), 568–579.

Chapter 10

Method for Simple and Routine Three-Dimensional Cell Culture

Eleanor Knight[a,b] **and Stefan Przyborski**[a,b]

[a] *School of Biological and Biomedical Sciences, Durham University, South Road, Durham DH1 3LE, U.K.*
[b] *Reinnervate Limited, NETPark Incubator, Thomas Wright Way, Sedgefield, TS21 3FD, U.K.*
stefan.przyborski@durham.ac.uk

Growing cells using three-dimensional cell culture models is known to radically enhance cell functionality over conventional systems in which cells adapt abnormally to the two-dimensional growth substrate. This review article briefly considers some of the current technologies developed to enable three-dimensional cell growth in vitro and focuses on the practical issues surrounding their ability for routine use in cell culture laboratory. A simple solution for routine three-dimensional culture known as Alvetex® is highlighted, and methods for its use and application are described.

10.1 Introduction

It is well established that the growth of cells using traditional culture methods is far removed from the complexities cells experience in real-life tissues. One of the major physical differences relates

Cellular in vitro Testing: Methods and Protocols
Edited by John W. Haycock, Arti Ahluwalia, and John M. Wilkinson

ISBN 978-981-4364-97-3 (Hardcover), 978-981-4364-98-0 (eBook)
www.panstanford.com

to shape and geometry cells acquire when grown on a flat substrate as in a conventional cell culture plate or flask. Cells adapt to their local environment and growth on two-dimensional (2D) surfaces, resulting in cell flattening and remodeling of the internal cytoskeleton. Such changes have been shown to alter gene expression and thus in turn alter basic cell structure and function [1]. Accordingly existing 2D cell culture models are often a poor proxy for the growth and study of cells in vitro.

Over recent years there has been a gradual development and adoption of technology that enables cells to acquire their more natural shape and form. The application of this "three dimensional" (3D) cell culture is anticipated to enhance the physiological relevance of experiments performed in vitro. It is important, however, to clearly define what is meant by this term. 3D cell culture is about creating suitable surroundings for optimal cell growth, differentiation, and function by (i) allowing individual cells to maintain their normal 3D shape, structure, and function with minimal exogenous support and interference; (ii) encouraging cells to form complex interactions with adjacent cells and receive and transmit signals; (iii) enabling a more natural environment to foster the creation of the native architecture found in tissue structures; and (iv) reducing stress and artificial responses as a result of cell adaptation to flat, 2D growth surfaces. Various alternative methods have been developed to meet the growing demand for 3D cell culture. These can be broadly categorized into either scaffold-free or scaffold-based culture systems, with scaffolds made from either natural or synthetic materials.

Scaffold-free systems primarily comprise self-assembled hanging drops, resulting in the formation of microtissue spheroids. Although this technology has been used for many decades, it has more recently been developed into a system for routine screening applications [2]. The ability to precisely control the size of small cell aggregates is an advantage. In contrast, larger spheroid structures lose heterogeneity and suffer from necrotic centers [3]. Yet this can be a useful feature for modeling hypoxia in cancer research [4]. Furthermore, the ability to form layered and organized structures that more closely mimic the scale and ordered complexity of real tissues is limited using cell aggregate models.

The use of scaffold-based 3D culture models extends the range of options available to researchers. There are several different methods that can be broadly divided into two approaches, notably, hydrogels and solid scaffolds. The growth and suspension of cells in a hydrogel that comprises a cross-linked natural base material such as agarose, fibrin, collagen, or hyaluronic acid with high water content is a popular option [5]. Hydrogels can be designed to support specific types of cell growth and function by trapping cells in an artificial extracellular matrix protein environment that may be modified to incorporate biologically active molecules. Cells will often grow as small isolated aggregates within the gel, and the ability to form ordered layered structures is restricted. Practical issues, including expense, self-life, preparation steps, and batch-to-batch inconsistency, also limit their application for routine use in 3D cell culture.

Seeding cells into a scaffold that provides a 3D space in which cells can grow and occupy to create a natural 3D tissue-like structure is another widely used approach. Various types of scaffold material in a range of alternative formats and presentations have been developed. Natural substrates, such as seaweed-derived alginate, have been used to support 3D growth either by encapsulation [6] or fabrication into a macroporous scaffold [7]. This method allows the growth of cells as individual spherical masses; however, this is not homogeneous throughout the material, and the large thickness of the scaffold is not supportive of efficient mass transfer of oxygen and nutrients. Biodegradable materials such as poly(lactic acid) and poly(glycolic acid) and their copolymers poly(lactic-*co*-glycolic acid) have previously been developed for use in 3D cell growth [8]. Many such materials were originally developed for their ability to degrade and aid the integration of transplanted cells with host tissues. While a distinct advantage for tissue repair, biodegradability is often detrimental for in vitro studies where the scaffold slowly falls apart and releases by-products locally that can subsequently influence cell behavior. Moreover, such materials are not practical for routine 3D cell culture where issues such as shelf life, storage, and product consistency need to be taken into consideration.

Inert nondegradable materials overcome these issues and can be engineered into highly porous materials suitable for 3D cell culture. Such scaffolds can consist of voids and interconnects or fibers and filaments, creating a large internal volume and space that cells can occupy. There are many methods by which such materials can be made, for example, electrospinning [9], gas foaming technology [10], leachable particles [11], and emulsion templating [12, 13]. Many variations of these approaches have led to small-scale advances in materials development and application to cell biology, but very few have been developed into a commercially successful process to consistently manufacture scaffolds for biological applications in vitro. This is with exception, however, for emulsion templating, a long-established technology that has recently been optimized to produce Alvetex® polystyrene-based scaffolds that have subsequently been commercialized for use in 3D cell culture [14–16].

Alvetex® is a highly porous, cross-linked polystyrene scaffold, which has been sectioned into 200 μm thick membranes (Fig. 10.1). It has been researched and developed as a cost-effective solution for simple, routine 3D cell culture with a high level of flexibility and versatility to satisfy the varying needs of the user. Polystyrene is an attractive substrate for 3D cell culture since it is familiar to the user, it has been tried and tested for many years, it is inert and does not degrade during normal use, it is compatible with the majority of downstream analytical methods, and it is a consumable product with no shelf life issues. The scaffold has been adapted to fit a variety of conventional cell culture plastic-ware formats. Careful consideration has been given to the presentation of the scaffold for alternative uses. For example, placement of the scaffold at the bottom of a culture vessel such as a multiwelled plate bathes the 3D culture with the medium from above only (Fig. 10.2A), whereas presentation of the scaffold in a well insert enables medium support from above and below the culture (Fig. 10.2B,C). These apparently subtle differences impact significantly the behavior of cells in the culture, and these alternative formats are best suited for short-term and long-term 3D culture, respectively. Alvetex® provides a suitable 3D structure in which cells can proliferate, migrate, differentiate, and function in an appropriate niche environment. As a result cells maintain a 3D shape and form close interactions with adjacent cells (Fig. 10.3).

Alvetex® technology has been designed for simple, routine 3D cell culture. The product is supplied sterile and is ready for immediate use. The following short methods provide some examples of how

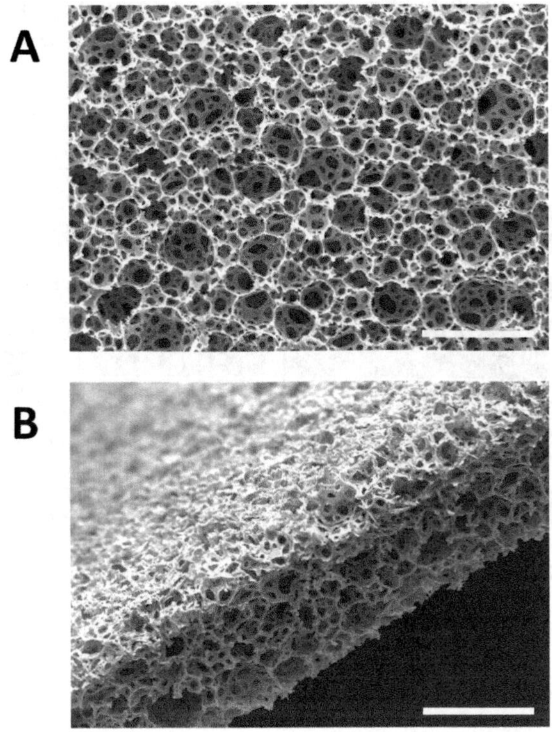

Figure 10.1 Structure of the Alvetex® polystyrene scaffold. Scanning electron microscopy reveals the highly porous nature of Alvetex®, which consists of voids and interconnecting pores. Micrograph shows (A) the homogeneous structure of the material. Alvetex® is engineered into a 200 μm thick membrane, as shown in transverse section in panel (B). (Scale bar = 200 μm.)

to prepare and use Alvetex® technology. Visit www.reinnervate.com for information about additional protocols and use of downstream analytical methods compatible with Alvetex® technology.

10.2 Materials

10.2.1 *Choosing the Right Alvetex® Format*

1. Alvetex® is available in a variety of formats, each specifically designed for different 3D cell culture applications: In the 12-well

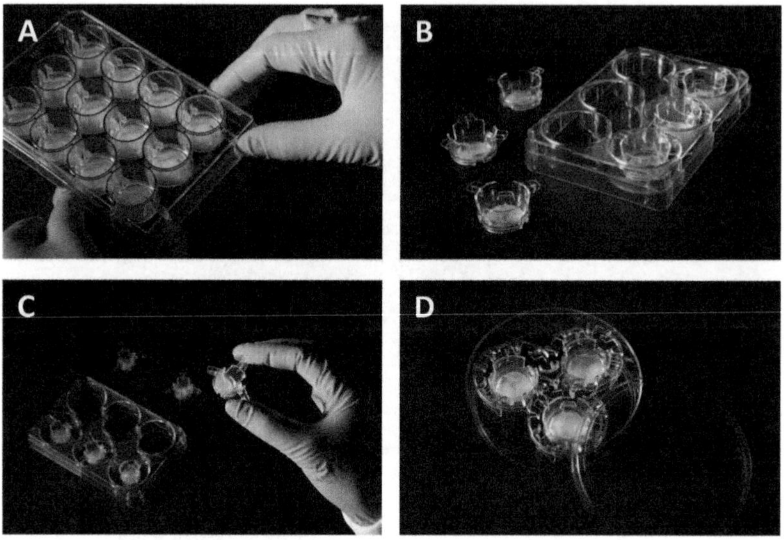

Figure 10.2 Photographs showing alternative formats of Alvetex® and associated devices used to present the scaffold in different ways. (A) 12-well plate, (B) 6-well insert, (C) 12-well insert fitted into a 6-well plate, and (D) a well insert holder fitted inside a petri dish containing three 6-well inserts. Note that the 12-well insert has long arms that can be snapped back for fitment into a standard 12-well plate.

plate (AVP002) it is located at the bottom of each well, whereas in well inserts (AVP004-3 and AVP005-3) Alvetex® is suspended. The well inserts are easy to use and are highly versatile. They fit a range of culture plates from different manufacturers as well as a custom-made well insert holder in a deep petri dish (AVP015).

2. In deciding which Alvetex® format to use, the following factors should be considered in combination: (i) cell type and duration of culture, (ii) desired depth of cell penetration in the 3D cell culture; and (iii) type of assay to be performed or application to be used. The rate of cell growth in Alvetex® is affected by cell metabolism, proliferation rate, motility, and cell size. It is therefore important to choose the Alvetex® format that will achieve optimal growth for your chosen cell type. For example, Fig. 10.4 shows two different cell types grown for the same

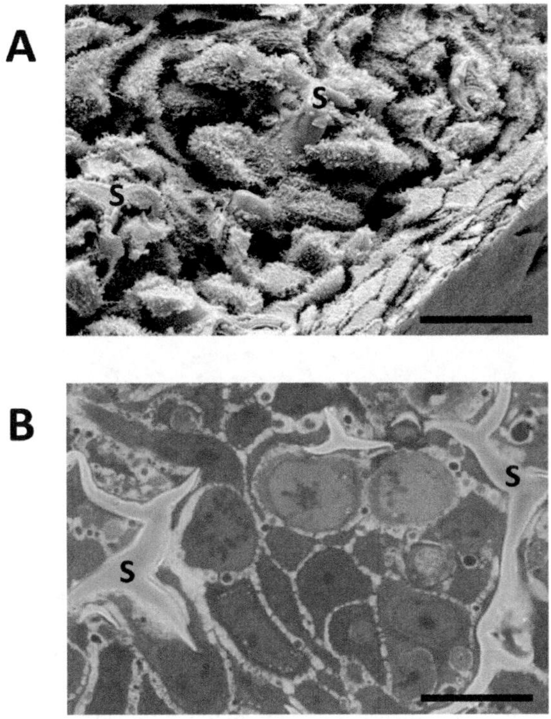

Figure 10.3 Cells maintain a 3D shape and form complex interactions with adjacent cells when grown in Alvetex®. (A) Scanning electron micrograph of HaCaT cells inside and on the surface of the scaffold (s). (B) Toludine Blue resin section showing histological details of the same cells growing in the scaffold. Scale bars = (A) 25 μm; (B) 10 μm.

time period using the same format of Alvetex®, resulting in two cultures with different characteristics.

3. Alvetex® formats have been specifically designed to enable optimal 3D cell culture in both short- and long-term experiments. The 12-well plate format is ideal for short-term cell cultures where the medium is replaced every one to two days. The Alvetex® membrane sits at the bottom of each well, and the cells are exposed to the medium only from above. This is desirable for studies where easy access is required to the cells, as they predominantly reside in the top portion of the scaffold (e.g., for transfection studies). Alternatively a shorter experiment in

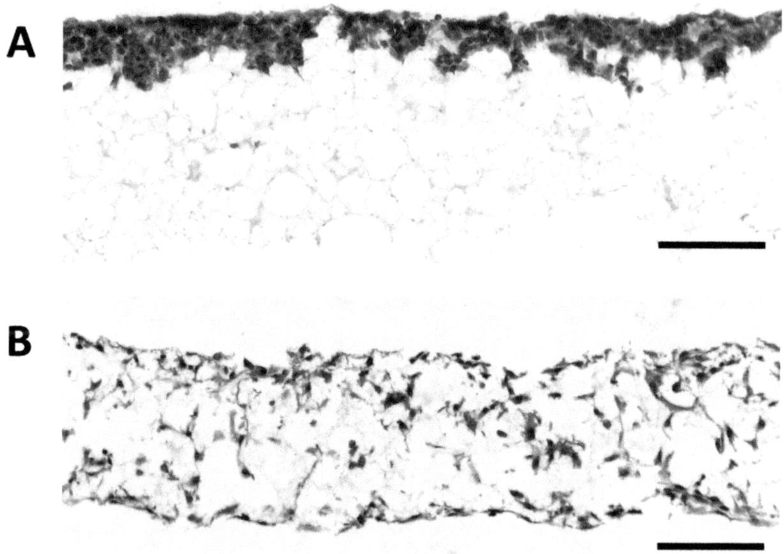

Figure 10.4 Not all cells types grow the same in a 3D culture. Cells differ in size, motility, metabolic requirements, and proliferation rate. The images show two different cell types grown on Alvetex® in the 12-well plate format for seven days. (A) HepG2 cells and (B) 3T3 fibroblasts. Note how the liver cells occupy the top 25–30% of the scaffold densely, whereas the fibroblasts penetrate the full thickness of the material in a more dispersed pattern. (Scale bar = 100 μm.)

well inserts is also suitable, as the cells will not have had time to penetrate the scaffold. The well inserts, on the other hand, are capable of supporting cell growth for up to three weeks for assays and applications where higher cell numbers are desirable. The design of the well insert allows for greater cell penetration into the scaffold, and generation of higher cell yields is achieved because the Alvetex® is suspended in the medium such that cells receive nutrients from above and below. Therefore, cells sustain optimal growth for longer and achieve greater differentiation, creating a cultured tissue that more closely resembles the growth of cells in the body. Well inserts and the well insert holder device can also be used to create coculture systems either in the same insert or in the insert and well combination. Table 10.1 provides

Table 10.1 Suggested guidelines for the use of Alvetex® formats for cell applications and assays

Application	12-well plate	6-well insert	12-well insert
Histology	■■■	■■■	■■■
Immunocytochemistry	■■■	■■■	■■■
Confocal microscopy	■■■	■■■	■■■
Light microscopy*	■	■	■
Viability assay	■■	■■■	■■■
Toxicity assay	■■	■■■	■■■
Proliferation assay	■■	■■■	■■■
Metabolic activity assay	■■	■■■	■■■
Gene expression**	■■	■■■	■■■
Protein expression**	■■	■■■	■■■
Growth at air–liquid interface	N/A	■■■	■■■
Cell-signaling assay	■■	■■■	■■■
Permeability assay	N/A	■■■	■■■
Transfection of 3D cells	■■■	■■	■■
Coculture	■	■■■	■■■
Invasion assay	■	■■■	■■■
Migration assay	■	■■■	■■■

■■■ most suitable
■■ suitable
■ least suitable
N/A, not applicable.
Ranking is based on Alvetex® disc format suitability, the likely cell yields, and therefore signal generation and whether exogenously added chemicals/cells can be contained to only one side of the membrane.
*The growth of cells cannot be followed by traditional light microscopy as in 2D, but similar to an ex vivo tissue, 3D structures have to be evaluated using histology or confocal microscopy. Alternatively cell proliferation can be monitored using a viability assay such as 3-[4,5-dimethylthiazol-2-yl]-2,5 diphenyltetrazolium bromide (MTT). In addition, cells can be visualized and viewed under a conventional light microscope subsequent to staining with neutral red stain (see www.reinnervate.com).
**Nucleic acids and protein can be extracted directly from cells in 3D culture using standard cell lysis and extraction protocols.

guidance for the selection of alternative formats in line with the application or downstream analytical procedure required.

4. The well insert holder in a deep petri dish (AVP015) is designed to hold well inserts in a large-volume petri dish, reducing the need for frequent medium changes (Fig. 10.2D). This format should be used for prolonged cell growth of highly proliferative and demanding cell types. The well inserts can be positioned at

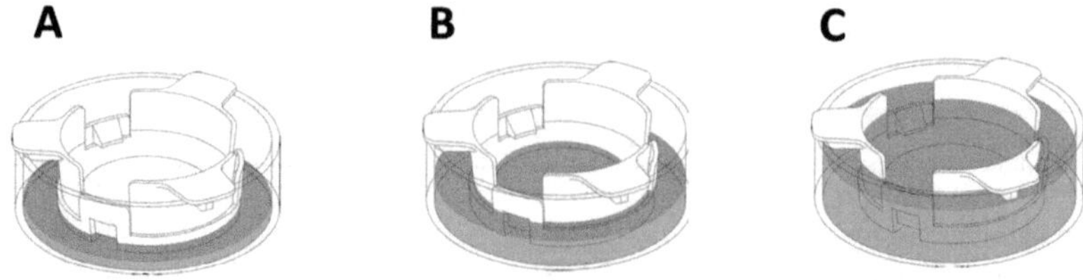

Figure 10.5 Well inserts designed for flexibility. The well insert houses Alvetex® in its base and comprises polystyrene walls with three equally spaced openings toward the top of the wall. This allows for different medium fill levels suitable for alternative applications. (A) Medium in contact with the 3D culture from below only—3D growth at the air–liquid interface. (B) Medium in contact above and below but independent compartments—3D growth with two different types of medium. (C) Medium in contact above and below but compartments connected through the opening in the wall—long-term 3D growth with equal medium composition on either side of the Alvetex® membrane.

three different levels in the insert holder to allow for cultures to be raised to the air–liquid interface (for air–liquid interface differentiation) and subsequent permeability/barrier testing (Fig. 10.5). 3D cocultures can also be set up in one or two of the well inserts within the same petri dish.

5. Overall these products are designed to provide cell biologists with a broad range of choice and great flexibility when designing their 3D cell culture experiments.

10.3 Methods

10.3.1 *Handling Alvetex®*

1. All procedures concerning the handling of Alvetex® should be performed wearing gloves according to standard aseptic methods required for cell culture in a class I/II cabinet.
2. When dry, Alvetex® is reasonably fragile with a wafer-like consistency; however, once rehydrated the discs become more robust. Therefore handle the material carefully when performing any manipulation, including medium changes, transferring the discs for analysis, fixing and embedding for histology, etc. When

using forceps, exercise care as manipulating the scaffold can damage its structure. Try to handle the Alvetex® discs around the edges only.

3. When dispensing liquids (e.g., 70% ethanol [EtOH], phosphate buffered saline [PBS], and the medium) over Alvetex®, place the end of the pipette tip toward the wall of the culture vessel, avoiding touching the scaffold. If using the 12-well plate format, retain the cylindrical clip in place. Well inserts can be fed from the outside: place the end of the pipette tip toward the wall of the culture vessel (going either through the window of the well insert or beside it), let the liquid rise gently to touch the base of the well insert, and, if required, dispense the rest of the solution into the well insert to prevent it from floating. Seed cells on the middle of the disc without touching the membrane itself.

10.3.2 *Use of Alvetex® in 12-Well Plate Format*

The 12-well plate format is a simple presentation of Alvetex® technology: it comprises a single loose disc and clip per well. The clip holds Alvetex® in position during transit and use. The clip is made from polystyrene, it is sterile and inert, and it can easily be removed to release the Alvetex® disc for subsequent analysis of the cultured cells.

10.3.2.1 Preparing nontreated Alvetex® for use and cell seeding

1. Open the 12-well plate carefully to ensure that the clips holding the Alvetex® discs are not displaced.
2. Add approximately 2 mL of 70% EtOH to each well to pretreat the Alvetex® disc in preparation for incubation in aqueous solutions (e.g., PBS, culture medium).
3. Carefully aspirate the EtOH solution and immediately wash the Alvetex® disc in ~2–3 mL of an appropriate medium for ~1 min.
4. Carefully aspirate the medium wash and replace with the final wash medium (use the same type of medium as for cell seeding). The Alvetex® disc is now ready for cell seeding: aspirate the

medium just before application of cells. If preparation of cell suspension is delayed, incubate the plate with the medium at 37°C with 5% CO_2 until further use.
5. Prior to cell seeding, Alvetex® can also be precoated with standard cell culture reagents such as collagen, fibronectin, laminin, poly-D/L-lysine, poly-L-ornithine, and Matrigel to encourage cell adhesion, differentiation, and function optimization. Perform this step after EtOH treatment and appropriate buffer wash steps instead of the medium.

10.3.2.2 Optimization of seeding and 3D cell culture using the Alvetex® 12-well format

The 3D cell culture is different from the conventional 2D cell culture and as such requires optimization according to cell type:

1. For most applications initial cell-seeding densities of $0.5 - 2.0\times 10^6$ cells in 100–150 μL per disc are recommended. Seeding in a low volume enables cells to attach predominantly to the Alvetex® disc and avoids cell loss on other surfaces.
2. When inoculating, aspirate the washing medium thoroughly from the plate and carefully dispense cells on the middle of the discs. Replace the lid and incubate in a humidified incubator at 37°C with 5% CO_2 for 15–30 min to facilitate cell attachment.
3. After this time gently flood the wells with the medium by dispensing up to 4 mL of the medium per well.
4. With 3D cell culture there will be many more cells growing per unit volume of the medium. Therefore, users must refresh the mediums more frequently, ideally once every one to two days; however, this will also depend on the population-doubling rate and nutrient demands of the cell type cultured.

10.3.3 *Use of Alvetex® in 6-Well Insert and 12-Well Insert Formats*

The presentation of Alvetex® in well insert formats is versatile, enabling long-term 3D culture as cells can receive nutrients from the mediums above and below the membrane, sustaining optimal 3D cell growth. Two sizes will be demonstrated (Fig. 10.2): AVP004-3 (22 mm diameter) and AVP005-3 (15 mm diameter). Both are supplied in blister packs with three individually sealed inserts containing Alvetex®. The 6- and 12-well inserts are designed to fit into most 6-well plates or the custom-made "Well Insert Holder in Deep Petri Dish" (AVP015). Snapping the extended wings of AVP005-3 will also enable it to fit into a 12-well plate.

10.3.3.1 Preparing nontreated Alvetex® (6- and 12-well inserts) for use and cell seeding

1. Immersion in 70% EtOH will instantly pretreat Alvetex® in preparation for incubation in aqueous solutions (e.g., PBS, culture medium). This can be done by dipping the well insert into a beaker containing 70% EtOH before placing it into the chosen holder vessel.
2. Alternatively EtOH treatment can be performed in situ once the well insert is positioned in the plate. Add sufficient 70% EtOH to the well so that the level of the liquid rises above the membrane (for 6-well plates add approximately 5 mL/well; for 12-well plates add approximately 2 mL/well).
3. Carefully aspirate to waste and immediately wash Alvetex® in an appropriate medium (for 6-well plates use 7 mL/well; for 12-well plates use 2.5 mL/well) for ~1 min.
4. Aspirate and replace with the final wash medium (use the same type of medium as for cell seeding). The scaffold is now ready for cell seeding: aspirate the medium just before application of cells. If preparation of cell suspension is delayed, incubate the plate with the medium at 37°C with 5% CO_2 until further use.
5. Prior to cell seeding, Alvetex® in the well insert can also be precoated with standard cell culture reagents such as collagen, fibronectin, laminin, poly-D/L-lysine, poly-L-ornithine, and Matrigel to encourage cell adhesion, differentiation, and function optimization. Perform this step after the EtOH treatment, followed by an appropriate buffer wash step instead of the medium.

10.3.3.2 Optimization of seeding and 3D cell culture using Alvetex® 6- and 12-well insert formats

The 3D cell culture is different from the conventional 2D cell culture and as such requires optimization according to cell type, the assay being performed, and the insert configuration used (Fig. 10.5).

1. For most applications initial cell-seeding densities of 0.5–2.0 $\times 10^6$ cells in 100–150 μL per disc in a 6-well insert or 0.25–1.0 $\times 10^6$ cells in 50–75 μL per disc in a 12-well insert are suitable. Seeding in a low volume enables cells to attach predominantly to the disc and avoids cell loss on other surfaces.
2. When inoculating, aspirate the washing medium thoroughly from the plate and carefully dispense cells on the middle of the discs without touching the membrane. Replace the lid and incubate in a humidified incubator at 37°C with 5% CO_2 for 30 min to 90 min to facilitate cell attachment.
3. After this time gently flood the wells with the medium by dispensing 3.0 – 10.5 mL of the medium per well in a 6-well plate or if using a 12-well insert in a 12-well plate dispense 1.4–4.0 mL of medium per well. Fill up the wells carefully beside the insert so that the medium comes up from the bottom to gently contact the cellularized Alvetex® disc and gradually floods the insert itself.
4. The volume of the medium required will depend on user requirements, and recommendations are outlined in Table 10.2.

Table 10.2 Recommended feeding volumes for different well insert configurations

	Feeding volumes for alternative fill levels*		
Well insert and plate type	Below only	Above and below separately	Above and below interconnected
6-well insert in a 6-well plate	3.5 ± 0.5 mL/well	7 ± 1 mL/well	10 ± 0.5 mL/well
12-well insert in a 6-well plate	3.5 ± 0.5 mL/well	7 ± 1 mL/well	10 ± 0.5 mL/well
12-well insert in a 12-well plate	1.6 ± 0.2 mL/well	2.4 ±0.2 mL/well	3.8 ± 0.2 mL/well

*See Fig. 10.5 for reference.

10.4 Notes

1. **In 3D cell culture there will be more cells per unit volume of medium.** Therefore, users must refresh the medium more frequently, typically every 2±1 days; however, this will also depend on the population doubling rate, nutrient demands of the cell type cultured, and the volume of the medium used, as described above. If any sign of cell attachment and growth is evident on the bottom of the plate, transfer the well inserts into a new plate, refeed, and then incubate as usual.
2. **Use of a well insert holder in a deep petri dish.** A deep petri dish enables users to grow their 3D cultures in larger volumes of mediums compared to an ordinary multiwell plate. Up to 95 mL of mediums can be used in a deep Petri dish, and it is therefore capable of sustaining long-term 3D culture experiments (three to four weeks) and reducing the frequency of medium exchanges. If required, a magnetic stirrer bar can be placed at the bottom of the dish to circulate mediums and facilitate exchange. A single well insert holder is capable of housing up to three well inserts (6- or 12-well inserts; Fig. 10.2) in a deep petri dish. A well insert can be positioned at three different levels in the insert holder: high, medium, and low (Fig. 10.6). This feature allows cultures to be raised to the air–liquid interface by moving the insert to a different level within the same holder. Positioning the well inserts at different levels may also be used to conserve expensive mediums or allow for increasing medium volumes for demanding cell types over the course of a long-term experiment. This is also in combination with altering the level of the medium in and around the well insert, as depicted in Fig. 10.5. Table 10.3 provides information concerning the different volumes of mediums required for alternative setups.
3. **Coculture setups.** The well insert holder system also allows for the 3D coculture of more than one cell type by seeding different cells in one or two of the well inserts within the same petri dish. 3D cocultures can also be set up within the same well insert. Alternatively, a support cell line can be cultured at the base of the petri dish in 2D and another in 3D culture in the well insert. It is important to ensure that suitable mediums are chosen

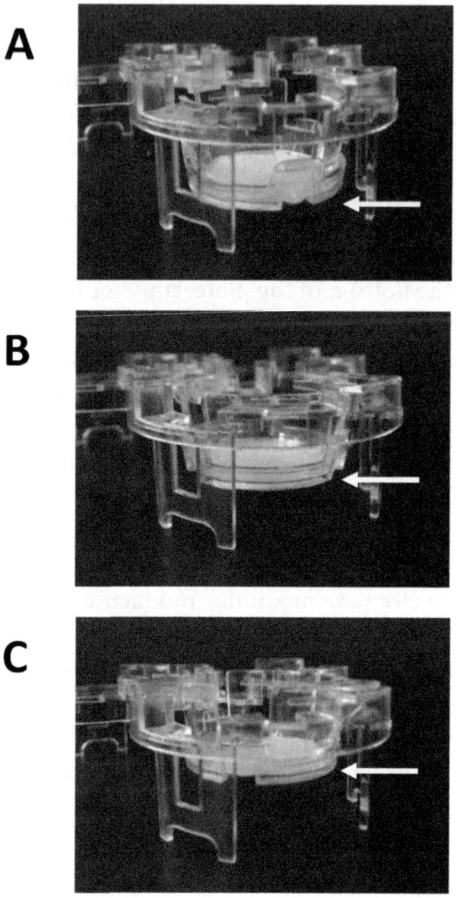

Figure 10.6 Photographs showing three alternative positions for a well insert in a holder designed to fit inside a petri dish. Arrow indicates the base of a 6-well insert at (A) low, (B) medium, and (C) high locations.

that will simultaneously support the growth of both cell types cultured. The well insert holder will fit into deep petri dishes with approximate dimensions of 86 mm (internal diameter) ×25 mm (height).

4. **Comparison of 3D cell growth patterns on the Alvetex® 12-well plate format**. Alvetex® in the 12-well plate format was treated with EtOH and washed with the complete medium

A

B

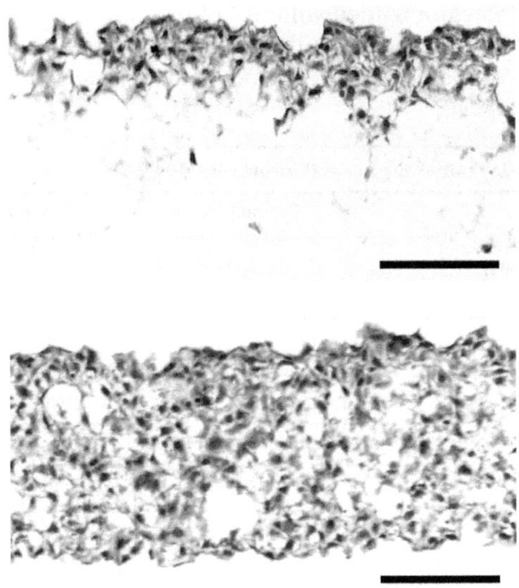

Figure 10.7 The presentation of the scaffold affects the growth pattern of the culture. In this example, the same number of HaCaT cells are grown for seven days as submerged cultures in either a 12-well plate (A) or a 6-well insert (B). Note how these cells penetrate and grow much more extensively when the culture is also supported by the medium from below the membrane as well as above (B). (Scale bar = 100 μm.)

prior to cell seeding (the complete medium consisted of DMEM, 10% FBS, 2 mM L-glutamine, and 100 U/mL penicillin and streptomycin). HaCaT cells (a human keratinocyte cell line) were plated at a density of 0.5×10^6 cells in 150 μL per well, while HepG2 cells (a human liver cell line) were seeded at a density of 2×10^6 cells in 150 μL per well. Plates were incubated for 30 min before flooding with further mediums and maintained for 7 days. After preserving in Bouins fixative the discs were paraffin-embedded, sectioned (10 μm), and counterstained with hematoxylin and eosin. Fibroblast cultures demonstrated significant cell invasion into the matrix, while HepG2 cells remained resident in the top 25% of the matrix (Fig. 10.4). These data show how alternative cell types behave differently in 3D culture, which is an important factor to consider during experimental design.

Table 10.3 Recommended volumes of medium required when positioning either the 6-well or the 12-well insert at different levels in the holder. Three additional combinations are included for when using the well insert at different fill levels

Recommended volumes for a 6-well insert (AVP004-3)			
	Feeding volumes for alternative fill levels*		
Well insert setting within holder	Below only	Above and below separately	Above and below interconnected
Low	20 ± 1 mL/well	40 ± 3 mL/well	70 ± 5 mL/well
Medium	34 ± 2 mL/well	50 ± 3 mL/well	80 ± 3 mL/well
High	48 ± 2 mL/well	70 ± 5 mL/well	92 ± 3 mL/well
Recommended volumes for a 12-well insert (AVP005-3)			
	Feeding volumes for alternative fill levels*		
Well insert setting within holder	Below only	Above and below separately	Above and below interconnected
Low	20 ± 1 mL/well	40 ± 3 mL/well	72 ± 5 mL/well
Medium	34 ± 2 mL/well	52 ± 3 mL/well	82 ± 3 mL/well
High	48 ± 2 mL/well	70 ± 5 mL/well	92 ± 3 mL/well

*See Figs. 10.5 and 10.6 for reference

5. **Comparison of 3D cell growth patterns of HaCaT cells on Alvetex® presented in various formats.** HaCaT cells were seeded (0.5×10^6 cells in 150 μL per well) on EtOH-treated and complete medium–washed Alvetex® scaffolds in the following formats: a 12-well plate (AVP002), 6-well inserts (AVP004-3) in a 6-well plate, and 12-well inserts (AVP005-3) in a well insert holder in a deep petri dish (AVP015). Cultures were maintained for seven days in a complete medium. After preserving in Bouins fixative the discs were paraffin-embedded, sectioned (10 μm), and counterstained with hematoxylin and eosin. In general, significantly more proliferation and cell invasion into the scaffold are observed in 3D cultures grown in a well insert (medium interconnected feeding regime) (Fig. 10.7). This is due to nutrient availability from above and below the scaffold and is particularly noticeable when well inserts are used in combination with the well insert holder in a deep petri dish, resulting in the formation of a slab of tissue-like material.

References

1. Vergani, L., Grattarola, M., Nicolini C. (2004) Modifications of chromatin structure and gene expression following induced alterations of cellular shape. *Int J Biochem Cell Biol* **36**, 144–1461.
2. Kelm, J. M., Lorber, V., Snedeker, J. G., Schmidt, D., Broggini-Tenzer, A., Weisstanner, M., Odermatt, B., Mol, A., Znd, G., Hoerstrup, S. P. (2010) A novel concept for scaffold-free vessel tissue engineering: self-assembly of microtissue building blocks. *J Biotechnol* **148**, 4–55.
3. Yoshii, Y., Waki, A., Yoshida, K., Kakezuka, A., Kobayashi, M., Namiki, H., Kuroda, Y., Kiyono, Y., Yoshii, H., Furukawa, T., Asai, T., Okazawa, H., Gelovani, J. G., Fujibayashi, Y. (2011) The use of nanoimprinted scaffolds as 3D culture models to facilitate spontaneous tumor cell migration and well-regulated spheroid formation. *Biomaterials* **32**, 6052–6058.
4. Hirschhaeuser, F., Menne, H., Dittfeld, C., West, J., Mueller-Klieser, W., Kunz-Schughart, L. A. (2010) Multicellular tumor spheroids: an underestimated tool is catching up again. *J Biotechnol* **148**, 3–15.
5. Tibbitt, M. W., Anseth, K. S. (2009) Hydrogels as extracellular matrix mimics for 3D cell culture. *Biotechnol Bioeng* **103**, 65–663.
6. Zimmermann, H., Shirley, S. G., Zimmermann, U. (2007) Alginate-based encapsulation of cells: past, present, and future. *Curr Diab Rep* **7**, 314–320.
7. Dvir-Ginzberg, M., Elkayam, T., Cohen, S. (2008) Induced differentiation and maturation of newborn liver cells into functional hepatic tissue in macroporous alginate scaffolds. *FASEB J* **22**, 144–1449.
8. Mikos, A. G., Sarakinos, G., Leite, S. M., Vacanti, J. P., Langer, R. (1993) Laminated three-dimensional biodegradable foams for use in tissue engineering. *Biomaterials* **14**, 323–330.
9. Sun, T., Mai, S., Norton, D., Haycock, J. W., Ryan, A. J., Mac Neil, S. (2005) Self-organisation of skin cells in 3D-electrospun polystyrene scaffolds. *Tissue Eng* **11**, 1023–1033.
10. Salerno, A., Oliviero, M., Di Maio, E., Iannace, S., Netti, P. A. (2009) Design of porous polymeric scaffolds by gas foaming of heterogeneous blends. *J Mater Sci Mater Med* **20**, 2043–2051.
11. Aydin, H. M., El Haj, A. J., Piskin, E., Yang, Y. (2009) Improving pore interconnectivity in polymeric scaffolds for tissue engineering. *J Tissue Eng Regen Med* **3**, 470–476.
12. Bokhari, M., Carnachan, R. J., Przyborski, S. A., Cameron, N. R. (2007a) Emulsion-templated porous polymers as scaffolds for three

dimensional cell culture: effect of synthesis parameters on scaffold formation and homogeneity. *J Mater Chem* **17**, 4088–4094.

13. Carnachan, R. J., Bokhari, M., Przyborski, S. A., Cameron, N. R. (2006) Tailoring the morphology of emulsion-templated porous polymers. *Soft Matter* **2**, 608–616.
14. Bokhari, M., Carnachan, R. J., Cameron, N. R., Przyborski, S. A. (2007b) Culture of HepG2 liver cells on three dimensional polystyrene scaffolds enhance cell structure and function during toxicological challenge. *J Anat* **211**, 567–576.
15. Bokhari, M., Carnachan, R. J., Cameron, N. R., Przyborski, S. A. (2007c) Novel cell culture device enabling three-dimensional cell growth and improved cell function. *Biochem Biophys Res Commun* **354**, 1095–1100.
16. Schutte, M., Fox, B., Baradez, M. O., Devonshire, A., Minguez, J., Bokhari, M., Przyborski, S., Marshall, D. (2011) Rat primary hepatocytes show enhanced performance and sensitivity to acetaminophen during three-dimensional culture on a polystyrene scaffold designed for routine use. *Assay Drug Dev Technol* **9**, 475–486.

Index

For Product Safety Concerns and Information please contact our EU representative GPSR@taylorandfrancis.com Taylor & Francis Verlag GmbH, Kaufingerstraße 24, 80331 München, Germany

Batch number: 10725937

Printed by Printforce, the Netherlands